Reframing Convenience Food

Peter Jackson • Helene Brembeck
Jonathan Everts • Maria Fuentes
Bente Halkier • Frej Daniel Hertz
Angela Meah • Valerie Viehoff
Christine Wenzl

Reframing Convenience Food

SUSFOOD

palgrave
macmillan

Peter Jackson
Department of Geography
University of Sheffield
Sheffield, UK

Helene Brembeck
Centre for Consumer Science
University of Gothenburg
Gothenburg, Sweden

Jonathan Everts
Institute of Geosciences and Geography
Martin-Luther-University
Halle-Wittenberg, Germany

Maria Fuentes
Centre for Consumer Science
University of Gothenburg
Gothenburg, Sweden

Bente Halkier
Department of Sociology
Copenhagen University
Copenhagen, Denmark

Frej Daniel Hertz
Department of Communication and Arts
Roskilde University
Roskilde, Denmark

Angela Meah
Department of Geography
University of Sheffield
Sheffield, UK

Valerie Viehoff
Institute of Education
University College London
London, UK

Christine Wenzl
Institute of Geography
University of Bonn
Bonn, Germany

ISBN 978-3-030-08626-8 ISBN 978-3-319-78151-8 (eBook)
https://doi.org/10.1007/978-3-319-78151-8

Cover illustration: Jochen Tack / Alamy Stock Photo

Printed on acid-free paper

This Palgrave Macmillan imprint is published by the registered company Springer International Publishing AG part of Springer Nature.
The registered company address is: Gewerbestrasse 11, 6330 Cham, Switzerland

Preface and Acknowledgements

What do we mean when we refer to convenience food? How do consumers make sense of the category? How do they combine it with other kinds of food and modes of cooking? How do they incorporate it into their everyday lives and what are its consequences in terms of public health and environmental sustainability? These are some of the questions that this book seeks to answer, challenging received wisdom, taken-for-granted ideas and common-sense assumptions about the topic. By rooting our understanding of convenience food in the mundane practices of everyday life, we seek to reframe the subject, questioning how the popularity and commercial success of convenience food can be reconciled with its conventional positioning as unhealthy and unsustainable. How to resolve this paradox lies at the heart of our suggested reframing of convenience food.

The research on which the book is based was funded under the FP7 ERA-Net SUSFOOD programme on sustainable food production and consumption (FP7-291766). The SUSFOOD programme aims to reinforce cooperation in research, development and innovation between EU members and associated States in order to maximize the contribution of research to the development of more sustainable food systems. The book was written during the Brexit process, when the UK was negotiating its departure from the EU. Working together has, however, only strengthened the authors' common commitment to collaborative working within and beyond the European Research Area. We are also happy to acknowledge our gratitude to Nikola Schulz, the SUSFOOD project manager, to Annika Fuchs at the Federal Agency of Agriculture and Food (BLE) in Germany, and to our European funding partners: the Federal Ministry of Food and Agriculture in Germany (BMEL), the Danish Research Council (DASTI), the Department for Food, Environment and Rural Affairs (Defra) in the UK and the Swedish Research Council (FORMAS).

The book reports on four case studies (commercial baby food, supermarket ready meals, workplace canteen food and home-delivered meal boxes) all of which offer ease and convenience to consumers at different points in the cycle of planning, shopping, cooking, eating and clearing up. The research was undertaken in Denmark, Germany, Sweden and the UK, by a multi-disciplinary research team with expertise in anthropology, geography, sociology and communications research. Our aim has been to work across the four countries and to integrate the case studies. Individual chapters may focus on one country and one kind of convenience food (such as supermarket ready meals in the UK or meal-box schemes in Denmark). But each chapter is organized thematically and includes comparative material from across the four cases with the aim of producing a genuinely integrated, international and interdisciplinary study.

Each chapter was initially drafted by one or two authors and they are credited here: the original version of Chap. 1 was drafted by Peter Jackson, Chap. 2 by Peter Jackson and Jonathan Everts, Chap. 3 by Bente Halkier, Chap. 4 by Helene Brembeck and Maria Fuentes, Chap. 5 by Jonathan Everts and Peter Jackson, Chap. 6 by Christine Wenzl and Jonathan Everts, Chap. 7 by Bente Halkier and Angela Meah, Chap. 8 by Angela Meah and Maria Fuentes, and Chaps. 9 and 10 by Peter Jackson. But, in each case, other authors contributed to successive drafts so that every chapter is multi-authored and the book is a truly collective endeavour.

Our collaboration was greatly enhanced by periodic writing workshops in Gothenburg, Mallorca and Sheffield where successive chapters were read and discussed, arguments were refined and new examples added prior to redrafting and recirculation. We also presented our work at several international meetings, sharing preliminary thoughts with each other, with invited discussants and wider audiences. The following meetings were particularly productive and we would like to thank the organizers for creating the opportunity for open dialogue and lively debate: the Nordic Conference on Consumer Research meetings in Vaasa (2014) and Aarhus (2016) and the European Sociological Association meetings in Prague (2015) and Bologna (2016).

Finally, we would like to thank our home universities and academic departments for providing a stimulating and supportive environment for our research. These include: the Department of Geography at the University of Sheffield where Peter Jackson and Angela Meah are based; the Centre for Consumer Science (CFK) at the University of Gothenburg where Helene Brembeck and Marie Fuentes are based; the Department of Communication at Roskilde University where Bente Halkier began the project and where Frej Daniel Hertz's doctoral research is based; the Department of Sociology at Copenhagen University to which Bente Halkier transferred in 2016; and the Department of Geography at the University of Bonn when Jonathan Everts began the project, before moving to Dresden and then to Halle Universities. Valerie Viehoff was also employed in Bonn while working in the UK and Germany, and Christine Wenzl's PhD research is based there. It is a measure of the openness of our academic communities and the porousness of disciplinary boundaries that several of us spent time at each other's institutions, sharing ideas and deepening our understanding of our four countries and their relationship with (different kinds of) convenience food. We are grateful for such collegiality and for all the opportunities we have been afforded in the process of working together. Our families also deserve our special thanks for enabling us to work on this project, including days spent away from home and hours working late. Thanks for the many ways you have enriched our lives and for making our work on this book more convenient.

Sheffield, UK	Peter Jackson
Gothenburg, Sweden	Helene Brembeck
Halle-Wittenberg, Germany	Jonathan Everts
Gothenburg, Sweden	Maria Fuentes
Copenhagen, Denmark	Bente Halkier
Roskilde, Denmark	Frej Daniel Hertz

Sheffield, UK	Angela Meah
London, UK	Valerie Viehoff
Bonn, Germany	Christine Wenzl

Contents

List of Figures

1

Introduction

Convenience food is a complex and contested category, whether understood as a marketing term used by retail professionals or in everyday conversation by non-specialists. It encompasses a wide variety of processed and semi-processed foods including frozen pizza and ready meals, sausages, sandwiches and pies, tinned fruit and canned vegetables, bagged salads, confectionary and crisps—all of which might broadly be described as 'ready to cook, ready to heat or ready to eat' (see Pfau and Saba 2009).

Despite these definitional issues (pursued in more detail in Chap. 3), convenience food is frequently criticised as unhealthy and environmentally unsustainable, responsible for eroding the distinctiveness of local food-ways as part of a wider process of cultural homogenization, sometimes referred to as McDonaldization (Ritzer 1993). For example, a study published in the *British Medical Journal* reported that none of the 100 ready meals it tested conformed to minimum WHO dietary standards (Howard et al. 2012),[1] while a report from the UK Department for Environment, Food and Rural Affairs criticised convenience food for the inclusion of resource-intensive ingredients with high greenhouse gas emissions and heavy transport costs, consuming large volumes of energy, land and water (Defra 2012). In these circumstances, the use of convenience

P. Jackson et al., *Reframing Convenience Food*,
https://doi.org/10.1007/978-3-319-78151-8_1

food is often 'tinged with moral disapprobation' (Warde 1999, p. 518), particularly in comparison with 'home-made' food, cooked from scratch using fresh ingredients.

This book and the research project on which it is based questions this pejorative view, seeking to understand how various types of convenience food have become embedded in consumers' lives, combined with other kinds of food to become part of their everyday diet. Seeking to understand the place of convenience food in people's daily lives, we ask what lessons can be learned from the commercial success of convenience food that might be applied by those who seek to promote healthier and more sustainable diets? The project draws on original findings from comparative research in Denmark, Germany, Sweden and the UK, funded through the ERA-Net SUSFOOD programme. The book argues that reframing convenience food within an understanding of everyday consumer practice provides new academic insights and helps avoid the all-too-frequent moralization of convenience food in policy and media circles.[2]

This introductory chapter outlines the complexity and contested nature of convenience food (cf. Jackson and Viehoff 2016). It introduces the research project on which the book is based including the rationale for our four case studies (commercial baby food, supermarket ready meals, workplace canteen food and meal-box schemes). It describes our theoretical and methodological approach, concluding with an outline of the book's core argument.

Convenience Food as a Contested Category

Convenience food is a sprawling category that defies easy definition (Scholliers 2015). Given the lack of an agreed definition (explored further in Chap. 3), we could take various approaches to the topic. One would be to adopt the definitions used by our research participants, following the term wherever it takes us, as suggested by Gluck and Tsing's (2009) approach to tracing 'words-in-motion'. Alternatively, we might try to delimit the field and focus on specific types of convenience food (such as ready meals or frozen pizza). Our approach is somewhat different from either of these perspectives. Rather than seeking to arbitrate whether some kinds of food should be included in the term while others should not, we have sought to trace how specific foods in particular circumstances come to be regarded as convenient (a process we describe using the neologism 'conveniencization'). How, for example, does powdered baby milk (infant formula) come to be regarded as more convenient than breast milk? Where and when does this happen? Whose interests does it serve and whose does it marginalize or exclude? What commercial forces enable it to happen and what socio-technical innovations are involved in its development? Rather than taking 'convenience food' as a separate category of food whose meaning is settled and unchanging, we examine the processes and practices through which certain foods take on characteristics that are regarded (by some people in some places and at specific times) as 'convenient'.

Our focus on 'conveniencization' (explained in more detail in Chap. 3) also enables us to explore how new forms of convenience food become normalized parts of people's diets, regarded as staples within the practices of their everyday lives (cf. Lavelle et al. 2016). It also encourages us to explore how many households are able to combine 'fresh' and 'convenience' foods without making a strong distinction between the two categories (cf. Carrigan and Szmigin 2006; Short 2006). The process of 'conveniencization'—and the distinction it implies between 'convenient' and 'convenience' food—is crucial for understanding what we mean by 'reframing' convenience food—and what is at stake in approaching the subject in this way.

The FOCAS Project

The research on which this book is based was funded by the ERA-Net SUSFOOD programme whose aim was to enhance collaboration and coordination between European research programmes on sustainable food production and consumption. The programme involved 25 partners from 16 European member states. The Call to which we responded addressed three thematic areas (with our project being associated with the third strand): increased resource efficiency in food production; innovation in food processing technologies; and understanding consumer behaviour to encourage more sustainable food choice.

Our project addressed the relationship between Food, Convenience and Sustainability (or FOCAS, for short). Its specific aims were to examine:

- How 'convenience food' is understood by consumers and how its use relates to understandings of 'healthy eating' and environmental sustainability
- With what specific practices (shopping, cooking, eating, disposing) 'convenience food' is associated
- How such foods are incorporated within different household contexts and domestic routines, and
- To what extent current practices are subject to change (towards more sustainable and healthier practices)?

We addressed these aims through four carefully-chosen case studies, each addressing different aspects of convenience food. The first case explored the consumption of **commercial baby food** among families with children aged under 18 months. Following debates on food safety, sustainability and health (Bentley 2014), the understanding of processed baby food as a modern, healthy and scientifically-based product has been questioned. This was particularly true in China, where the contamination of infant formula with the poisonous chemical melamine led to widespread parental anxieties about baby milk and led the dairy industry to adapt their marketing strategies (cf. Gong and Jackson 2012, 2013). Today, processed baby foods are marketed as a convenient and flexible solution for time-pressed mothers, meeting their needs while conforming to official health advice. However, elaborate packaging and long-distance transport, complex supply chains and changing dietary guidelines make the choice of baby food a vexing issue. Our research (led by Helene Brembeck and Maria Fuentes) was based in Sweden with some comparative work elsewhere in Europe including brands such as Semper in Sweden, Hipp in Germany, Ella's Kitchen in the UK and Lovemade in Denmark. It included desk-based research on the marketing of various brands of baby food and field-based research among Swedish and Somali women in the small Swedish town of Falköping, exploring the role of processed food in baby-weaning practices as mothers sought to introduce their children to solid food. This case study allowed us to test Alan Warde's (1999) claim that baby food is rarely marketed in terms of its convenience for babies or mothers.

The second case study, based in the UK but including comparative fieldwork in Germany (led by Peter Jackson, Angela Meah and Valerie Viehoff), looked at an iconic example of convenience food: **supermarket ready meals**, sold in frozen or chilled form, at numerous price points and with separate branded and own-label options. The research used interviews, kitchen 'go-alongs' and ethnographic observation with a diverse range of households in the UK and Germany to examine how consumers incorporate a range of convenience foods within their diets; the meanings attached to their consumption; and the scope for introducing healthier and more sustainable alternatives. We examined, in particular, the relationship between convenience and care (Meah and Jackson 2017) and associated ideas about food and family life, including the negative moralization of convenience food (explored in more depth in Chap. 7).

The third case, based on fieldwork in Germany (led by Jonathan Everts and Christine Wenzl), examines the communal provision of **workplace canteen food** where customers forego an element of choice in exchange for having their meals cooked for them. This case study began from the premise that many consumption 'choices' are not made by sovereign individuals but by institutional actors working on behalf of individual consumers. Constrained food choices are common in workplace canteens and other institutional settings where the range of food options may be quite restricted. In such cases, individual customers exert limited influence over the choice of ingredients, the sourcing of food or the way it is prepared. In these circumstances, the meaning of 'convenience' is likely to vary, as when eating in a canteen is regarded as less demanding than bringing food from home. While institutional catering may provide opportunities for 'choice editing' in support of health and sustainability agendas, it also reduces consumer choice (especially for those with 'minority' food preferences such as vegetarians or those who follow a halal or kosher diet).[3] Canteen food also offers potential benefits in terms of collective procurement with possible advantages in terms of environmental sustainability (less packaging and waste, less reliance on individual journeys to shop etc.). The research for this case study was based in Euskirchen, a small town in a largely rural area, and Düren, a larger industrial town in the North Rhine Westphalia region of Germany. Participatory observation involving informal conversations, 'eat alongs' and shared coffee breaks were held with guests and staff in eight canteen kitchens and dining rooms including a courthouse, two manufacturing companies, a vocational training centre, a hospital, bank, army base and town hall. Interviews, averaging 20–40 minutes, were conducted with three canteen managers, three chefs, three assistant cooks, a trainee and an intern. Interviews with guests were shorter, averaging 10–15 minutes, aiming to cover a wide range of social groups, including variations by gender, occupation and work tasks, religion and ethnicity.

The final case study (led by Bente Halkier and Frej Daniel Hertz) looked at **meal-box schemes** in Denmark. The research examined two specific schemes—*Ret Nemt* ('Fairly Easy') and *Årstiderne* ('The Seasons')—which made different claims regarding their healthiness and sustainability in terms of their use of organic and/or local products. This case study allowed us to test whether some degree of convenience (home delivery and carefully measured amounts of food suitable for specific recipes) was compatible with ideas of healthy home-cooking. Combining interviews, focus groups, participant observation and media reception analysis, around a dozen families with school-age children were studied in Denmark's Zealand Region.[4]

While all of the case studies focused on a specific country, each had a comparative element, more developed in some cases than in others. We also made a specific decision to study small towns and provincial cities in contrast to the overwhelming metropolitan focus of existing food research (cf. Atkins et al. 2007; Franck 2005). While some might challenge our reliance on case studies in terms of the difficulty of generalizing from a narrow sample, we follow Mitchell (1983) in arguing that case and situation analysis depends on a different logic from statistical generalization (where large numbers of cases are required). Rather than relying on *statistical* inference, we deploy what Mitchell describes as *logical* inference where a single well-chosen case study can be more effective than large numbers of badly-drawn examples. Likewise, our selection of cases does not depend on a 'sampling' strategy where the aim is to reduce bias by identifying a representative sample. Rather, we deploy a process of maximum variation sampling, aiming for a diversity of cases and maximizing the diversity of participants within each case study (Flyjbjerg 2001) as part of a constant comparative method (explained further in Chap. 3). Finally, the study was strongly interdisciplinary, drawing insights from anthropology, geography, sociology and communications research.

A 'Theories of Practice' Approach

The book adopts a 'theories of practice' approach to convenience food, drawing inspiration from a wide range of sources, as outlined by Reckwitz (2002) and as discussed in relation to consumption studies by Warde (2005). Reckwitz defines a practice as 'a routinized type of behaviour which consists of several elements, interconnected to one another: forms of bodily activities, mental activities, 'things' and their use, a background knowledge in the form of understanding, know-how, states of emotion and motivational knowledge (2002, p. 249).[5] Specifically, we follow the work of Theodore Schatzki who argues that practices provide a conceptual middle ground between individual action and social order (Schatzki 2002). From this perspective, the unit of analysis is the practice—not individuals (who are the carriers of those practices) or structures (which only exist insofar as they are reproduced through practices). For Schatzki, practices are at the centre of the social world, constituting the 'site of the social' in conjunction with the material arrangements amidst and through which practices transpire. According to Schatzki (1996, p. 86), practices involve a nexus of 'doings and sayings', opening up a space for the observation of social practices as well as participants' discursive reflections on their practices. Practice theory also encourages us to examine how specific practices such as cooking and eating are negotiated within a wider set of domestic routines and responsibilities (cf. Wills et al. 2015; Warde 2016).

By employing a practice theory perspective, we emphasize that food is a process rather than a stable entity. Food is produced, packaged, distributed, bought, brought home, opened, prepared, cooked, served, eaten, digested and so on. Food has no meaning outside the practices and bodily processes of food production and consumption. It is for this reason that we chose to turn to practice theories when considering the categories of convenience food and convenient foods. By resorting to practice theories, we acknowledge the importance of what people do and say. From this perspective, social phenomena such as convenience food are best understood as an outcome as well as an ingredient of social practices. In our case, and as stated above, we are interested not so much in the formal category of *convenience food* but in the multiple ways that different kinds of food are made *convenient*, through the process of *conveniencization* (see Chap. 3 for an elaboration of these terms).

Practice theories have been applied to a range of fields by Warde (2005), Shove and Pantzar (2005), Jackson and Everts (2010) and Halkier et al. (2011). In relation to food, early applications considered the practices of shopping for food (Jackson et al. 2006; Everts and Jackson 2009; Blake et al. 2010; Meah and Watson 2013). Practice theories have also been used in research on domestic kitchens and dining rooms (Wills et al. 2015), food provisioning (Jackson 2009), cooking (Meah and Jackson 2013), eating (Warde 2016), disposing of food (Evans 2012) and handling media-contested food routines (Halkier 2010).

Drawing on practice theory and with a focus on time pressures, time shifting and the competing demands on modern household schedules, Alan Warde (1999) recognised convenience food as a thought-provoking topic for social scientific analysis, while Jackson and Viehoff (2016) suggested that practice theories might help open up the study of convenience food for a more nuanced understanding of their uses and meanings in the context of everyday life.

Practice theory provides us with a common vocabulary and set of concepts to put to work in our case studies, emphasising the *meanings* that consumers attach to convenience food; the *practices* with which convenience food is associated; the *embeddedness* of convenience foods in the routines and rhythms of everyday life; and the '*do-ability*' of convenience food in terms of its technical feasibility and cultural appropriateness (Halkier 2010). Practice theory also helps understand how consumers are recruited to specific practices associated with the use of convenience food; how such practices become routinized and habitual; how they are supported by specific forms of socio-technical apparatus and infrastructure; and how they involve particular kinds of skill and competence.

An Outline of the Book

Following this introduction, Chap. 2 traces the historical growth of consumer demand for various types of convenience food, acknowledging the significance of earlier forms of bottled, pickled and canned food but focusing on the period beginning in the 1950s with the development of the frozen TV dinner in the United States and contemporary European examples (including frozen, chilled and ambient products, branded and own-label). It discusses the variable market penetration of convenience food across Europe and examines the role of technological change including innovations in industrial food processing (such as the 'cold chain') and domestic technologies (such as refrigeration, home freezing and microwave cooking). The chapter also considers the role of supermarkets in shaping the routines of car-borne food shopping and changing gender relations and household structures (including the effects of increased female participation in the labour force and the growth of single-person households). The chapter ends with a more detailed account of the development of convenience food in Denmark, Germany, Sweden and the UK.

Chapter 3 discusses the nature of convenience food as a complex and contested category, subject to multiple interpretations and diverse uses (cf. Halkier 2013). It explores the difficulties of translating 'convenience food' into other European languages besides English and how the category is used to refer, variously, to fast foods, snack foods and packaged/canned/frozen/pre-prepared foods. The chapter shows that convenience food is a contested category among academic researchers (who employ the concept as an analytical term) and among consumers (who use the term in everyday life). Drawing a distinction between 'convenience' food as a marketing category that applies to certain kinds of food and a wider range of foods that are rendered 'convenient' through specific consumer practices, the chapter advances our core concept of 'conveniencization' to refer to the process through which certain kinds of foods come to be recognised as more or less convenient than others. The chapter proposes an initial comparison of our four case studies in terms of heating, cooking and eating practices, examining what, where and when foods are rendered convenient. We also propose a typology of how foods become convenient in terms of their acquisition, appropriation and appreciation.

Chapter 4 demonstrates how convenience foods have become incorporated within people's everyday routines and dietary practices including the way shopping and cooking convenience foods have been normalized. Examining historical and contemporary sources, the chapter shows how commercial baby food in Sweden has needed continuous work to reconcile its use with notions of being a 'good mother'. The chapter shows how certain practices associated with processed baby food are scripted (involving notions of prescription, de-inscription and re-inscription) in relation to advice from health authorities and other official bodies, or in response to marketing campaigns. Commercially-produced baby food is considered convenient in enabling parents (usually mothers) to feed their children in a variety of locations, at home and 'on the move'. While it is often regarded as an acceptable and modern way of infant feeding, based on ideologies of 'scientific motherhood', it can pose significant problems in terms of cultural appropriateness, given competing (idealized and highly gendered) ideologies about 'feeding the family'.

Chapter 5 outlines the complex temporalities of convenience food including long-term historical changes in technology and society, generational and life-course changes, and shorter-term changes such as treats and rewards. Drawing particularly on the ready meals case study, the chapter shows how convenience food has been employed to help resolve issues of scheduling and routinization, through time-saving and time-shifting strategies. More abstract temporalities are considered including ideas of 'being modern' and the role of food in various kinds of memory-work. The chapter also considers the tactic of 'stocking up' in anticipation of future needs (where convenience food plays a key role). Having considered the temporal practices involved in shopping, cooking and eating, the chapter provides a more systematic comparison of the transformations, rhythms and timings associated with convenience food, concluding with a discussion of how convenience food is caught up in processes of escalation (doing more things) and acceleration (doing things within a shorter time).

Drawing particularly on the case study of workplace canteens as a form of collective food provisioning, Chap. 6 explores the spatial organization of convenience food. The chapter begins by considering the anxieties that are attached to the widening spaces of food production, distribution and retail, often described in terms of increasing 'food miles'. More complex spatialities are then described including spatial relations, places and spatial change, offering an alternative framing of convenience food. The specific spatialities of canteen food are then addressed in terms of networks and circulation, food preparation, and eating spaces.

Chapter 7 discusses the moralization of convenience food showing how its negative evaluation frequently involves implicit or explicit comparison with other sorts of food, using fresh ingredients, cooked 'from scratch'. The chapter demonstrates how convenience food is moralized through its associations with diet-related ill health, through deeply gendered ideas about maternal responsibility, and through arguments about the alleged decline of cooking skills (explored in more detail in Chap. 8). Convenience food is also caught up in contemporary debates about the responsibilization of consumers through notions of individualized 'food choice'. Our empirical evidence shows how participants justified their use of convenience food in relation to ideas about sustainability and waste, eating on a budget, and the need to accommodate family members' dietary tastes and preferences. Their frequent use of irony and self-deprecating humour highlights the moral ambivalence attached to convenience food. The chapter also challenges the common distinction between convenience and care, suggesting that the use of convenience food can be justified as an expression of care rather than as evidence of a dereliction of domestic duty.

Chapter 8 discusses the notion that the use of convenience food is associated with an alleged decline in cooking skills and culinary competence. Despite the popular 'discourse of decline' in media rhetoric and political debate, evidence for these assertions is actually rather limited, incomplete and out-dated. The chapter begins with some definitional issues, seeking to uncover what counts as 'cooking' in different contexts and how this has changed within living memory. Using the meal-box case study as the primary reference point, the chapter explores the skills associated with planned, improvised and audit-based approaches to cooking. It provides evidence of different forms of understanding including tacit knowledge, know-how and improvisation, concluding that meal-box schemes provide a convenient approach to meal planning while maintaining the positive values of home-cooking.

Chapter 9 reviews the implications of our study in terms of the health and sustainability of convenience food. It challenges the assumption that all convenience foods are unhealthy and unsustainable, countering deficit approaches which assume that consumers lack knowledge or skill regarding the food they consume. The chapter also challenges the view that consumers are unable to make connections between food and health or between food and the environment, exploring the reasons why consumer behaviour often departs from the 'best practice' encouraged by official advice on healthy eating or sustainable diets. Our research finds ample evidence of consumer interest in environmental questions, often expressed through the (academically contested) concept of 'food miles', as well as a general repugnance for wasting food (in contrast to the common assumption of consumer profligacy). The chapter also addresses the ethical trade-offs and practical compromises that surround the use of different kinds of convenience food.

The concluding chapter draws the argument together and offers suggestions for future research. It assesses the difference it makes to approach convenience food in terms of social practice theory and the insights that can be drawn from our comparative, international and interdisciplinary approach. The chapter highlights the analytical utility of the distinction we draw between 'convenience' and 'convenient' food and the value of our key concept of 'conveniencization'. Conclusions are provided at the level of the project as a whole and for each of our four cases, including a series of inferences about the relevance of our work for policy and practice. The chapter summarises the lessons learnt from our 'reframing' of convenience food through a social practice lens in terms of the health and environmental implications of current consumption practices and the potential for developing healthier and more sustainable alternatives.

Notes

1. Significantly, however, the same source (Howard et al. 2012) found that home-made meals, following recipes from TV chefs, were even less likely to meet WHO guidelines, challenging simple assumptions about the relative merits of 'home-made' and 'convenience' food.

2. This may be an appropriate point at which to note that our use of the term 'consumer' frames the subject in distinctive ways, denoting a particular, culturally, politically and historically-specific, relationship to the market (when buying or selling food and other goods). Where other words such as 'citizen', 'public' or 'parents' are more appropriate, these terms have been used instead. On the 'making' of the modern consumer, see Trentmann (2005).
3. The idea of 'choice editing' is discussed in an interview with Tim Lang in *The Guardian* ('Does the consumer really know best?', 25 October 2007).
4. Further details of the participants in each of the case studies are given in the Appendix.
5. Elsewhere, Shove et al. (2012) seek to understand the dynamics of social practice in terms of the interaction between meanings, materials and competences.

2

A Short History of Convenience Food

Introduction

This chapter traces the growth of consumer demand for various types of convenience food from the frozen TV dinner in 1950s America to contemporary European examples (including frozen, chilled and ambient products, branded and own-label). It discusses the variable market penetration of convenience food across Europe and examines the role of technological change including innovations in industrial food processing (such as the 'cold chain') and in domestic technologies (such as refrigeration and microwave cooking). The chapter also considers the role of supermarkets in shaping the routines of car-borne food shopping and changing gender relations and household structures (including the effects of increased female participation in the labour force and the growth of single-person households). The diversity of convenience food as a category in the marketing literature and in academic research will be discussed in more detail in Chap. 3.

P. Jackson et al., *Reframing Convenience Food*,
https://doi.org/10.1007/978-3-319-78151-8_2

The 'Prehistory' of Contemporary Convenience Food

As was argued in the Introduction, convenience food is a complex and contested category with a long history, spanning decades and perhaps even centuries if we take the development of canning, pickling and preserving into account as precursors of more recent kinds of convenience food such as frozen pizza or supermarket ready meals. In her recent history of processed baby food in the United States, for example, Amy Bentley traces the 'invention' of industrialized food products back to the 1930s when the Children's Bureau publication, *Infant Care*, first mentioned canned fruits and vegetables (2014, p. 17). But there is a much longer history of salted, dried, smoked, pickled, canned and preserved foods, many of which were invented to provide sustenance for the armed forces when in combat or on the move. So, for example, Nicolas Appert's experimental use of glass jars for preserving food took place during the Napoleonic Wars (1799–1815), the industrial-scale canning of pork and beans was prompted by the needs of the US army during the Civil War in the 1860s and, several decades later, during the First World War, the British army subsisted on tinned foods such as corned beef and Irish stew.

Some of the earliest forms of mass-produced convenience food such as condensed milk were already commercially available in the 1850s while canned goods, such as Heinz baked beans, were first sold in the UK in the 1880s. Nineteenth-century chemists such as Justus von Liebig were responsible for specific innovations, including the first patented recipes for infant formula (in the 1860s), while the double-seamed can which increased the shelf-life of food by means of an air-tight seal was invented in 1888 (as discussed by Shephard 2000).

Martin Bruegel provides a nuanced account of 'how the French learned to eat canned food' during the nineteenth and early-twentieth centuries (Bruegel 2002). Noting that canning technology was largely invented in France, he explores why it took almost a century before French consumers were convinced to use such foods. Describing their initial repugnance towards canned food, Bruegel also cites reasons of cost, taste and safety that slowed its widespread adoption in France. Eventually, it seems, it was the army and the public school system that served as the conduit for the large-scale uptake of canned food during the First World War as these institutions provided a 'school for taste' that eventually led cans to become a routine presence in French households (ibid., p. 125).[1]

American Precedents

It was in the 1950s, however, that what might now be recognised as modern-day convenience foods first began to appear. Following some early experiments with serving frozen meals on US airlines in the 1940s, the first branded 'TV dinner' was produced by C.A. Swanson & Sons in the 1950s, in packaging that resembled a TV set, served in a segmented aluminium tray, ready for heating in the oven. In a series of iconic advertisements, Swanson's appealed to the needs of modern housewives. In one such image (see Fig. 2.1), a well-dressed woman returns home late from a shopping trip ready to prepare a hot meal for her husband ('I'm late—but dinner won't be!'). While her husband relaxes in an armchair with his newspaper, the woman smilingly sets about making dinner. The clock in the background and the glance at her watch, emphasise the time pressures she faces, balancing her domestic duties with her other commitments. The illustrator, Lucia Lerner, produced a series of similar adverts in the 1950s with captions that reflect the dynamics of contemporary family life: 'When you're painted into a corner at dinner time', 'How to catch the early, early show with an easy, easy dinner', 'Extra guest for dinner? You're ready for him'. The advertising copy (text and imagery) promotes the wholesomeness of the meal: 'lean, tender slices of beef' with 'three, not two, vegetables', squaring the circle of convenience and care offering 'oven-quick meals that taste home-cooked', 'with no work before, or dishes after', described as a 'comforting thought for a hurried housewife'.

In a similar vein, Amy Bentley suggests that American women welcomed the invention of industrialized baby food as a symbol of modernity and convenience, consistent with the precepts of 'scientific motherhood' (Apple 1987; see also Chap. 4). Anxious to provide a nutritious diet for their babies, processed baby food offered mothers the

Fig. 2.1 Advertisement for Swanson's TV dinners. Source: Swanson, from https://www.flickr.com/photos/87362701@N00/234998824

freedom and mobility they craved while firms like Gerber, Heinz and Beech-Nut provided assurances about the nutritionally balanced nature of their products. Not until the 1960s were concerns raised about the amount of salt, sugar and other additives in commercial baby food, with campaigners like Ralph Nader and critics such as Marion Nestle heralding a consumer backlash against the power of the food corporations leading to an increase in breastfeeding, later introduction of solid foods and more reliance on 'natural' (home-prepared) foods. Bentley concludes that commercial baby food's great popularity 'was always largely about convenience', reducing preparation time and allowing mothers greater mobility because of the portability of glass jars and, later, pouches (2014, p. 148). More worryingly, Bentley suggests, commercial baby food helped prime the infant palate for the kind of industrialized products, high in salt, fat and sugar, that are now associated with the 'obesity epidemic':

> Commercial baby food was emblematic of mid-century American strength and power, complementing other such societal values as modernity and mobility. Its little jars of products laden with sugar, salt, and starch were gateway foods to the industrialized American diet … Using them made mothers feel confident and modern, fueling capitalist enterprise and quelling more 'natural' … alternatives, such as breastfeeding and homemade. (ibid., p. 163)

No less importantly, she concludes: 'Commercial baby food made it easier and more convenient for women with small children to enter the paid work force and stay there' (ibid., p. 163).

European Variations

The twentieth century saw a radical transformation of European kitchen spaces and cooking practices. The nineteenth-century kitchen was divided by class. Some bourgeois households could afford separate kitchens in the basement, operated by servants. For the rest of the population, coal or petroleum stoves within the living room (which was often the only room in the household) were the most common form of 'kitchen' (Oldenziel and Zachmann 2009). Over the course of the twentieth century, the kitchen became recast as the place for family life, community, identity and, most importantly, modernity. After the First World War, efforts intensified to design and produce affordable modern kitchens equipped with useful electrical appliances. This rush towards the modern kitchen, following the lead of the United States, was present in all European countries. The transition from coal stove to microwave cooking occurred roughly between the 1920s and 1990s. The transformation of the kitchen is crucial in understanding innovations and revolutions in food production and consumption such as the rise of convenience food. For convenience food relies on a stable network of electrical appliances that cool and warm food in convenient ways. This is true on either end of the production/consumption continuum and for all the places in between. In the kitchen, without fridges, freezers, electric stoves or microwaves, most convenience foods would not be convenient.

Despite the general modernisation of European kitchens, the adoption of convenience foods has been very uneven across Europe, particularly when specific kinds of food such as the ready meal are examined. In the UK, ready meals are the archetypal convenience food, their introduction having been spearheaded by major supermarket chains such as Marks & Spencer and Sainsbury's, following the invention of 'cold chain' technology in the 1950s. In other countries, such as Norway and Germany, frozen pizza occupies a similar place as the iconic convenience food, while in Mediterranean countries such as Italy, convenience foods have made much less impression (as noted in Carole Counihan's *Around the Tuscan Table*, 2004). Marianne Lien (1997) has documented the popularity of frozen pizza in Norway and the way one company responded to a market that was saturated with 'American' and 'Italian' brands by ingeniously introducing a local *folkepizzaen* (or people's pizza), designed to increase market share without cannibalizing existing brands.[2] Dale Southerton and colleagues have demonstrated the existence of significantly different consumption patterns in the UK and Spain, including variations in mealtimes and the prevalence of taking a mid-day meal at home (Southerton et al. 2011).

Significant variations exist even at a more local scale, within the Nordic countries. Based on a survey of >8000 consumers in Denmark, Finland,

Norway and Sweden and a comparative study of 800 consumers in St Petersburg, Kahma et al. (2016) conclude that attitudes towards convenience food were more positive in Norway than in the other Scandinavian countries, while almost half of respondents in St Petersburg reported never using convenience foods. Meanwhile, in a survey of over 22,000 Internet users, an AC Nielsen report found that UK consumers were among the most frequent users of 'ready to eat' (RTE) meals and that Danish consumers were the least likely to buy them, with 19% of Danes claiming never to purchase RTE meals and 49% claiming to buy them only seldomly. While nearly half of respondents globally would be unwilling to serve a ready-to-eat meal when entertaining friends or on a special occasion, two-thirds of French consumers reported their willingness to serve RTE meals when entertaining friends (AC Nielsen Global Report, Consumers and Ready to-eat Meals, December 2006).[3]

In what follows, we will explore in more detail the uneven historical moves towards convenience foods and convenient forms of meal preparation in the four countries of our case studies (the UK, Denmark, Germany and Sweden).

Convenience Food in the UK

The 1960s was the key decade in terms of the introduction of convenience foods in the UK. As well as the growth of supermarket shopping, innovations in domestic technology and an increase in female labour force participation, the 1960s saw the introduction of a range of iconic convenience food products. Named improbably after the Roman goddess of hearth and home, Vesta dried beef curry and chow mein were launched in 1961 and have been interpreted as evidence to the UK's growing passion for 'exotic' food (described in *The Guardian* newspaper as 'noshtalgia', 29 June 2002). Other products like instant mashed potato were also introduced in the 1960s, popularised by a 1974 TV commercial ('For mash get Smash'), where a group of loveable Martians look down on Earth and laugh at the way people still make mashed potato from raw ingredients, struggling to peel, boil and mash them by hand. Other products followed rapidly including, in 1967, Angel Delight: 'a chemically-laden powder whisked into a mousse with cold milk' (Mary Gwynn, 'A magical feast of memories…', *The Daily Mail*, 12 March 2015) which is still on sale today.

Food writer, Matthew Fort, recalls how 'When they first came out, ready meals were impossibly exotic and exciting, they were magical things. We all aspired to a Vesta curry' (quoted in Denise Winterman, 'The rise of the ready meal', *BBC News magazine*, 16 February 2013). Keen to develop products that combined convenience and quality, British food retailers sought to produce restaurant-standard food that would appeal to the increasingly sophisticated tastes of customers who had begun to take holidays abroad with the advent of budget air travel. At Marks & Spencer, for example, product developer Cathy Chapman was responsible for introducing chilled products such as Chicken Kiev and Chicken Cordon Bleu as 'respectable' alternatives to the frozen TV dinner. Food writer, Rose Prince recalls 'thinking it was rather posh … It seemed like something exotic and quite bistro' (quoted in Simon Usborne, 'Ready happy returns…', *The Independent*, 23 July 2009). The story is often told of how Cathy Chapman had to overcome Board members' resistance to the inclusion of garlic in their (now highly successful) Chicken Kiev dish when it was introduced in 1974. Chapman stood her ground and the product soon became a best-seller, rapidly followed by other dishes such as chilli con carne, Chinese spring rolls and chicken tikka masala (Kate Salter, 'Cathy Chapman: the woman who changed the way we eat', *The Telegraph*, 10 October 2010).

The appeal of ready meals in terms of the ease and speed of preparation can be inferred from the fact they were referred to in Marks & Spencer's in-house magazine as 'quickie meals' (*St Michael's News*, March 1984, quoted in Lin et al. 2017, p. 2). The market for ready meals expanded rapidly, driven by the major supermarket chains who all developed their own-label products as well as branded versions by companies such as Sharwood's. All of the major retailers now stock a large selection of ready meals, in frozen, chilled and ambient formats, including a wide range of 'ethnic' cuisines (such as Indian, Thai and Italian products) as well as traditional 'British' dishes such as shepherd's pie and 'bangers and mash' (sausage and mashed potatoes). It is often suggested that chicken tikka masala is now as much of a British staple as roast beef and Yorkshire

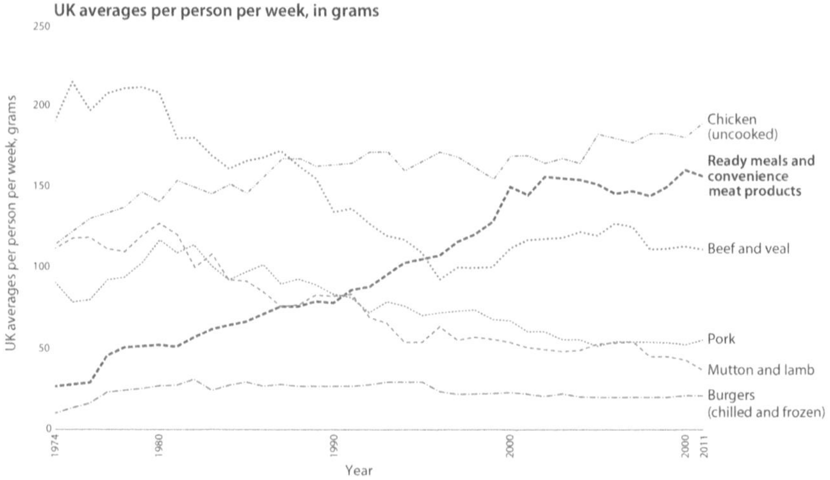

Fig. 2.2 Purchasing of meat products in the UK, 1974–2011. Source: Defra (2012)

pudding or fish and chips (see, for example, 'Fact file on nation's favourite dish', *The Telegraph*, 24 March 2004).

The rise of convenience food in the UK is relatively well documented. Figure 2.2 shows how the purchasing of ready meals and other meat-based convenience products has increased steadily since the 1970s while the consumption of beef, pork and lamb have declined. Data from the market research firm MINTEL, reported in the British press, claimed that UK consumers buy more ready meals than any other European country—double the level in France and Germany and six times higher than Spain ('UK tops ready meals consumption', *Daily Mail Online*: http://www.dailymail.co.uk/news/article-164609/UK-tops-readymeals-consumption.html, accessed 10 December 2016). While levels of demand for ready meals rose by 29% across Europe between 1998 and 2002, it grew by 44% in the UK. The report found that 30% of UK adults eat a ready meal more than once a week, compared with 16% in France.[4] UK consumers had also invested in the domestic technology needed to support their high level of ready-meal consumption with 80% of UK households owning a microwave oven compared to 27% in Italy. By 2012, the chilled ready-meal market was valued at £2.6 billion (BBC News magazine, 'The rise of the ready meal', 16 February 2013), projected to reach

£3.1 billion by 2017 (MINTEL, Prepared Meals, May 2013). More recently, an international study (reported in *The Guardian*, 3 February 2018) found that British people ate more 'ultra-processed' food, including salty snacks, sugary cereals, ready meals, cakes and cookies, than elsewhere in Europe. Such foods comprised 50.7% of the British diet, compared to 46.2% in Germany, 14.2% in France and 10.2% in Portugal. No data were reported for Denmark or Sweden.

As the market for ready meals has become increasingly saturated, new product development has begun to focus on premium brands such as Charlie Bigham's eponymous range of ready meals and a variety of healthier (low-fat, low-sugar) options such as Nestlé's Lean Cuisine and Sainsbury's Be Good to Yourself ranges. Marks & Spencer introduced 'nutritionally-balanced' ready-made meals for kids, 'based on Government guidelines for calories, fat, sugar and salt' ('Healthy eating for kids', *Your M&S* 2006, quoted in Lin et al. 2017, p. 40), followed by new product lines with names such as 'count on us', 'eat well' and 'balanced for you'. Supermarkets are also using 'celebrity chefs' to promote their products, such as Waitrose's Heston range (developed in association with Heston Blumenthal). Products are diversifying rapidly, offering several twists on the standard ready meal formula. One recent innovation is the provision of meal kits with pre-prepared ingredients in a single package, ready to cook in a few minutes. The artful combination of convenience (ease of preparation) with the comfort and care associated with home-cooking is signalled in one company's brand name 'Scratch', as shown in Fig. 2.3. Note, too, that the packaging makes a virtue of the nutritional value of the product and the lack of waste associated with having all the ingredients, precisely measured, chopped and ready to cook in one package ('healthy meals … nothing wasted').

There is much speculation about what drove these changes in the UK retail market and why the UK was such an early adopter of many kinds of convenience food. Commentators mention higher disposable incomes and increased time pressures resulting from the growing number of women entering the workforce; growing numbers of single-person households (which now account for one-third of UK households); longer working hours and an aging population, as well as the UK's increasing ethnic diversity.[5] It may also be relevant that, according to USDA

Fig. 2.3 Scratch meal kit. Source: © Sam Jackson

figures, UK consumers spend less of their household income on food (8.9%) compared to 11.4% in Germany and 13.4% in France (*The Telegraph*, 15 December 2017). But the role of high-street retailers such as Marks & Spencer, Tesco and Sainsbury's was also a key factor that differentiates the UK from other European countries with lower degrees of retail concentration.

Convenience Food in Denmark

Danish households began to experience a rise in wealth after the Second World War as a combined effect of economic growth, a steep rise in female employment and the building of the Danish welfare state (Buhl 2010). In 1960, women constituted 26% of the Danish workforce employed outside of the home, whereas in 2010 Danish women had an employment rate of 72% (Beskæftigelsesministeriet 2011). Although women were increasingly employed full-time, the main responsibility for providing household meals remained with women for a long period. This was still the case in 2006 (Deding et al. 2006) and only recently does research show that men's participation in shopping and cooking is beginning to grow significantly (Holm et al. 2016).

Also in the 1950s, new household food technologies such as the gas or electric stove, the refrigerator and the freezer were introduced alongside new types of retail stores such as supermarkets (Buhl 2010). All the new household food technologies were mass-produced industrial products, and the design of new kitchens was organised in order to be able to store and process foodstuff produced and distributed at large (industrial) scale. The new stoves became commonplace in Danish households in the 1950s, whereas refrigerators became common in the 1960s, and home freezers not until the 1970s (Thorndahl 2001). In this way, Denmark seemed to be later than the UK, Germany and Sweden in obtaining general household access to all of the new domestic technologies, necessary for taking full advantage of ready meals and similar convenience products.

The first self-service store with food was opened in 1949 by Brugsen (now Coop-Denmark), and the first supermarkets were established by the firm Irma (originally a chain of grocery stores), supported with funding from the Marshall Plan in order to increase productivity in the retail sector (Nielsen 2017). Supermarkets only became the dominant type of retail in Denmark at the beginning of the 1970s and Denmark got its first discount supermarket chain in 1977.

The provision of ready-made meals in Denmark did not begin until the 1970s and 1980s, and apparently did not seriously take off before the late 1990s, consistent with the results about Denmark being a comparative late-comer in terms of the consumption of convenience food in a European context (Nielsen 2006). The first commonly known and consumed type of ready meals were Chinese spring rolls from Daloon, a Danish firm established by a Chinese immigrant (Buhl 2010), and the next was 'The Saturday chicken', launched in 1982—a carved, marinated and spiced chicken in a tin-foil tray to put directly in the oven (Samvirke 2014). This ready meal became so iconic in the consumption of convenience food in Denmark that, even in the 2000s, some Danish consumers in interviews would refer to 'Saturday chicken' as a well-known food category (Halkier 2010, p. 147). By the end of the 1990s, a tendency with selling home-meal replacements in certain high-brow supermarket chains began (Gyldendal Encyclopaedia), and this has currently developed into a much greater variation in the supply and consumption of ready-made meals. The production and sale of 'The Saturday chicken' ceased in 2006.

The supply and sales of convenience products in Denmark has developed quite a lot in recent years. By the end of the 1990s, a tendency with selling home-meal replacements began, but at that time only in certain high-brow supermarket chains such as Irma in the Copenhagen area (Skibsted 2017). In general, the sale of ready meals has been growing in Danish retail, with the cool-chill solutions having the largest growth, compared to frozen ready meals and traditional tinned meals which are becoming less important. The most recent available statistics from the larger supermarket chains show a relatively significant growth over the last few years in the sales of convenience food products. The Dansk Supermarked chain sold 10–15% more ready meals in 2015 than in the previous year, together with 3% more meatballs and 7% more frozen pizza (*CPH Post*, 9 November 2015). The Danish Coop chain sold 26%

more convenience food products between 2013 and 2015, covering both ready meals and different processed foods such as chopped salads and cooked sauces (Coop-Analyse 2016). This seems to suggest that the formerly more hesitant and convenience late-comer consumers in Denmark are now joining the convenience trend more readily.

One way of understanding this development is to take account not only of the growth of the supply of convenience food products, but also importantly the greater diversity in the content of the products themselves. The Danish convenience food market has become much more varied over the past years, with a whole range of goods and meals of higher quality and healthier characteristics. This tendency for greater diversity in the market has also been spotted by marketing firms in the convenience food sector. One firm suggests that key trends for 2020 are a consumer emphasis on quality (in ingredients, materials and production methods) and convenience, seen as simplicity and ease of use (Nordic Food Survey 2015). Another review of the ready-meal market in Denmark suggests that Danish consumers' perception of ready meals as a good alternative to home-cooked dinners has slowly changed and that this can be explained by the diversification of the retail supply as manufacturers have increased their efforts to provide high-quality, innovative meals with good ingredients (Euromonitor International 2016a).

Looking more closely at the retail sector, the Coop supermarket chain in particular has taken some initiatives in their stores to boost the variety of convenience foods. In 2016, Coop launched their food manifesto, where one of the strategies was called 'Healthier to live easy'. This strategy builds on each Coop store having a 'Meal market' section close to the entrance, where chilled ready meals are presented. The products in the 'Meal market' section as well as a selection of other types of convenience foods such as salads, chopped vegetables and pizzas are labelled with the so called C-label. The C-label, according to Coop, stands for quality, freshness, taste and healthiness, and there are four different categories: ready to eat, ready to heat, ready for the salad bowl, and ready for cooking (Coop 2017).

Another recent trend in the convenience food sector in Denmark is internet-based meal-box schemes, where customers can order boxes with ready measured ingredients, delivered with recipes for specific meals. In

2017, Coop launched their own meal-box scheme as an add-on to their existing internet shop for home delivery of food, and thereby joined the two big players on the meal-box market, *Ret Nemt* ('Fairly easy') and *Åarstiderne* ('The Seasons').

Convenience Food in Germany

The German history is a bit more complicated than that of the UK or Denmark owing to its division into East Germany and West Germany after World War Two, up until reunification in 1990. In West Germany, the proliferation of supermarkets since the 1960s and the rise in female labour market participation (while retaining many domestic responsibilities, including the role of the main cook in the household) is similar to that of the UK and Scandinavian countries. In East Germany, the production and distribution of food was state planned and female labour force participation was much higher. Even in 2002, more than 10 years after reunification, East German mothers were three times as likely to be in full-time employment as their West German counterparts (Kreyenfeld and Geisler 2006). However, during the Cold War and beyond, in both parts of the country, cooking remained overwhelmingly the task of women. From the 1950s to the 1970s, the crucial era for the rise of convenience food, societal progress was not measured by who did the cooking but how the kitchen worked.

The modern kitchen was a political battleground during the middle third of the twentieth century. Mass consumption of fridges and electric stoves was the declared target of economic politics in Western and Eastern Germany. In the 1950s, the kitchen became the centre of a Cold War 'technopolitics' (Oldenziel and Zachmann 2009, p. 3).[6] During the American national exhibition in Moscow on 24 July 1959, the famous kitchen debate took place between Soviet Premier Nikita S. Khrushchev and American Vice-President Richard M. Nixon, starring the latest General Electric's kitchen design as background and point of reference for the two competing ideologies of capitalism and communism. As Oldenziel and Zachmann (2009, pp. 2–3) point out, the backdrop was crucial for the debate. In reference to another exhibited kitchen and its technological promises, Khrushchev allegedly said: 'They have no useful purpose. They are merely gadgets' (ibid., p. 9). However, the Soviet Union knew about the potential of refrigerators and other electric appliances and the desires they evoked.[7] Henceforth, technological innovations such as the fridge became a promoted product for mass consumption on both sides of the iron curtain, including West and East Germany.

The use of electrical equipment in German households remained low until the 1950s. The Nazis had campaigned for the fridge in 1936 in relation to pre-war-related strategies of reducing food waste (Heßler 2001, pp. 217–220). However, the fridge remained a luxury item available only for a few households. After the war, Ludwig Erhard, the minister of economy in West Germany, sought to revive the economy through campaigns aimed at increased household spending on consumer items. His first campaign in 1953, which was declared the 'year of the consumers', featured the fridge as the ideal product for the modern kitchen (Gerber 2015, p. 91). It was also favoured by energy companies, who lobbied for more steady consumption of electricity. As Gerber writes: 'the fridge as one of the electrical appliances with the highest steady electricity consumption in the kitchen became the symbol of progress, modernity, and progressive consumption in the emerging consumer society of Western Germany in the 1950s' (Gerber 2015, p. 92—our translation). Gerber further states that the 1950s were a transition period where thrift and increased consumer spending occurred alongside each other (ibid., p. 99). In the mid-1950s, Erhard made it clear that the fact that only one in twenty households possessed a fridge was a clear sign of backwardness, compared to the US or Switzerland, where fridges could be found in every second household. Although his campaigns for increased consumer spending faced some temporary backlashes, the overall picture is one of success. From 1950 to 1960, the percentage of households owning a fridge rose from a meagre 4 percent to an astonishing 40 percent (Gerber 2015, p. 99). In East Germany, the mass distribution of fridges started a little later but was well under way during the late 1960s (Wolle 1998, p. 28).

Following the mass consumption of fridges and freezers, the take-off of convenience foods in the 1960s and 1970s is well documented for West Germany:

> Freezers and frozen food had arrived for good in German households, of which more than half in 1978 had a freezer or freezing compartment. Fish fingers were followed in 1970 by the first frozen pizza produced in Germany; it was a huge success. That year, the average West German consumed 10 kg of frozen goods on top of ice cream, a number that was to quadruple during the following four decades. Modern convenience food had started with canned ravioli and the immensely popular Miracoli pack introduced by Kraft in 1961, a cardboard box that contained spaghetti, tomato sauce in a pouch, grated cheese and a herb-spice mix. Ready-made meals gained ground with the advent of microwave ovens, owned by one-third of West German households in 1989 (and virtually every single one could store frozen goods by then), a number that would double in the following three decades. (Heinzelmann 2014, pp. 317–318)

The history of convenience food in the GDR is more difficult to trace. It seems that classic West-German convenience food items such as the frozen pizza did not exist. It has been pointed out that the GDR's consumer culture was one of want (Wolle 1998). That does not mean that people were starving, on the contrary, there was enough food (Merkel 1998) but the state planned economy offered little choice and apparently little or no choice regarding convenience food.

Since 2005 there has been a significant decline in the number of people having lunch at home during weekdays. While the highest rates of home eating occur in childhood and later life, with lower rates between the ages of 16 and 60, these rates have all declined between 2005 and 2015 (see Fig. 2.4). Today, convenience food is the biggest trend in German food consumption. A recent report states that 55% of respondents want easy and quickly-prepared meals with 41% stating that they regularly eat frozen pizza or other ready meals. At the same time, people cook less, with only 39% cooking on a daily basis (BMEL 2017).

Convenience Food in Sweden

The development of convenience food in Sweden follows the American pattern, only one or two decades later, closely connected to the introduction of new technologies and increased female participation in the labour market.

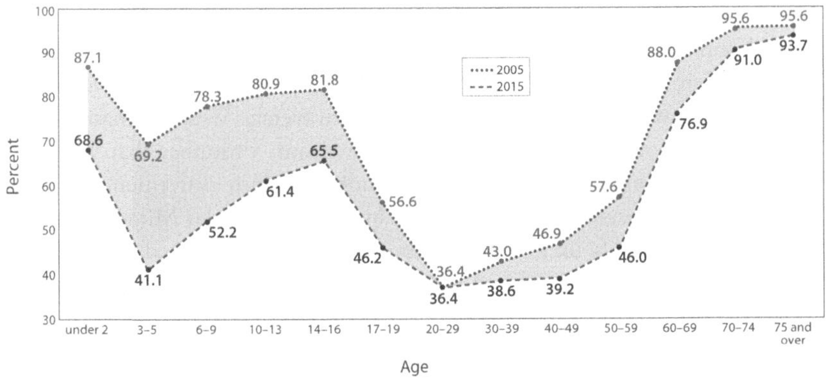

Fig. 2.4 Percentage of German people eating weekday lunch at home by age. Source: GfK ConsumerScan

During the 1920s the electric stove was introduced in Sweden, followed by domestic refrigerators which were to be found in almost every home by the 1950s (Hagberg and Kaijser 1987). Freezers (Sv. *frysbox*) were introduced in the 1950s and were commonplace by the 1960s. These technological developments were supported by political changes including, in 1944, the founding of the *Hemmets forskningsinstitut* (Home Research Institute) which fostered the idea of the home kitchen as a 'machine' including the use of freezers and microwave ovens for cooking frozen food and other convenience products (Hagberg and Kaijser 1987). Canned foods were first produced in 1869 by the Axel Molin Flinders cannery and were commonly available by the 1920s, promoted by the (then) canning company Findus in the 1940s with the slogan: 'Findus makes your household chores easier'.

The real expansion of convenience food is connected to the popularization of the freezer from the late 1950s. Felix launched a range of frozen foods such as ice cream, hash (*pytt i panna*), meatballs and French fries. But the main company connected with frozen foods in Sweden is Findus whose mission was to test new ideas from North America on the Swedish market (see Fig. 2.5). Findus started to produce frozen foods in 1945 and by 1950 they could be bought in 1500 stores.

The development of convenience food in Sweden is also closely connected to increased female participation in the labour force. In 1939 a law was passed that prohibited employers from dismissing women who

Fig. 2.5 Findus test lab, 1950s. Source: The Nordic Museum, Gullers Archives, © Peter Gullers

were pregnant or who planned to marry. Maternal and child benefits were introduced from the late 1940s and municipal nurseries and kindergartens began to be established. The largest increase in female employment occurred in the 1960s and 1970s, when nearly half a million women went out to work. Today, Sweden has the highest female employment rate in the EU.

In accordance with the Nordic welfare model, childcare is provided and subsidized by the state. 84% of children aged 1–5 are enrolled in nursery school in Sweden, while children in other EU countries generally start kindergarten at the age of 3–4. In Sweden, Finland and Denmark nearly 96% of six year-olds are in preschool. Today, the single-person household is the most common form of accommodation in Sweden. Over 1.5 million households consist of one person or almost 38% of all households. In an interview study with women born in the 1920s and 1930s (Brembeck et al. 2010) the women voiced how new packaging and products simplified their purchasing and cooking and were appreciated as expressions of modernity and progress:

> Previously, you had to buy milk from the dairy, and so on … it was so heavy to carry back home and took such an unbelievable amount of time to shop and cook and it was amazing when the frozen stuff arrived, and the semi-finished food was simply fantastic.

Some of the women now resented their enthusiasm for convenience food as young mothers:

> I belonged to the 'Findus mothers' then, who were eager to try all frozen. So it was *persiljejärpar* (mince-meat rolls with parsley) and fish sticks largely that fed my family' [she said laughingly, suggesting that this was obviously not the 'good' way to feed your family today].

By the late 1940s, Findus had a wide choice of frozen prepared foods from complete dinners in aluminum trays to various pastries. According to *The Findus Saga* (Findus 1962), celebrating the company's first 20 years, 'these products are not designed to replace meals prepared in the ordinary way, but to supplement them. Millions of people have neither the time nor facilities for full-scale cooking, and housewives welcome the availability of high quality foods which do not require many hours of kitchen preparation'.

The main problem with this line of product was how to combine, in one packet, foods which kept for different lengths of time and have different freezing points—combining fish or meat with vegetables and sauces so that every ingredient is fresh and palatable when prepared for the table (personal conversation with Torsten Andersson, chairman of Findus' Veterans Association, *Findus Veteranförening*). Frozen dinners never really took off in Sweden until the introduction of the microwave oven in the 1980s. Instead hospitals and canteens were the main customers of frozen dinners. The advantages of frozen dinners were marketed as 'exact portions, easy to prepare, saves time and labour' (Findus 1962).

The first self-service store opened in Stockholm in 1947, driving the development of convenience products. According to the women interviewed by Brembeck et al. (2010) they were considered fantastic inventions:

> and then suddenly there were these new stores called *snabbköp* (quick stores) in the 60s. You were supposed to go and pick the shelves yourselves.
> *Snabbköp* … it was hard to grasp … to simply go to the shelves and take the goods, and go to the cash register yourself.

During the 1960s and 1970s the quest for economies of scale and streamlining in food retailing drove the development of larger stores characterized by mass sales and high volumes, often in more peripheral locations, where the car was a necessity for consumers. Supermarket chains now began to appear more uniform across the country and the logos of the biggest actors, ICA and Konsum (now Coop), were introduced in the 1960s. In the 1980s, Findus started producing frozen single-portion tray meals and today Findus is the market leader in this sector, targetting single people in their twenties and older people in their seventies and above.

It is not, however, canned or frozen foods that are the iconic convenience food in Sweden, but the hot dog (*varm korv*). At the beginning of the twentieth century, the 'hot dog man' (*varmkorv gubbe*) selling hot dogs from a portable box filled with warm water was a common sight in Swedish cities (see Fig. 2.6). The traditional hot dog vendor disappeared for health reasons in the late 1950s, replaced by small kiosks that met hygiene requirements and there are still over 1500 hot-dog stands and snack bars (*gatukök*) in Sweden (Eriksson 2004).

The opening of the first McDonald's in Stockholm in 1973 was met with protests. McDonald's has since become 'normalized' in Sweden and a visit to one of the chain's restaurants is a common weekend practice for many families with children (Brembeck 2007). Convenience foods have come to play a significant part in Swedish family life, including the celebration of 'cozy Friday' (Brembeck 2012).[8] Today, Sweden has the most McDonald's restaurants in Europe and the hamburger can be said to be the iconic 'fast food', especially for young people, closely followed by pizza (Mäkelä et al. 2011)

Sushi came to Sweden in the 1970s and Tex-Mex in 1982 (Damberg 2015). Many other 'foreign foods' were introduced during the second half of the twentieth century. The first Chinese restaurant opened in Gothenburg in 1959 and the first Thai food restaurant in Stockholm in 1987. While frozen dinners are less popular in Sweden than in the UK,

Fig. 2.6 Hot dog vendor, 1950s. Source: © Lars Epstein

frozen products such as cabbage rolls, hash, meatballs and fish gratin were accepted and soon became normalized parts of the Swedish diet. The typical use of frozen convenience food is to use meatballs or chicken as one meal-ingredient in combination with potatoes and fresh vegetables. This is, of course, a good example of the 'conveniencization' of traditional home-cooked meals, elaborated further in Chap. 3.

Conclusion

The proliferation of convenience food and convenient cooking practices has a long history that can be traced back hundreds of years. However, the history that is most pertinent to convenience food items such as ready meals, frozen pizzas and the like belongs to the twentieth century. Technological innovations such as refrigerators, freezers and electric stoves and their adoption by virtually every household over the course of the century were crucial in providing the necessary domestic infrastructure. The mechanization and modernization of European kitchens did not just happen but were an explicit goal of economic policy, partially bound up in the great East-West antagonism of the Cold War era.

Many convenience products were introduced in Western Europe in the 1950s and became virtually ubiquitous in the 1960s and 1970s, connected with an increase in female labour force participation when time-pressed households needed quick and easy 'meal-solutions'. This explanation is only partial, however, as Eastern Europe, with a higher female labour force participation rate during the Cold War, lagged behind the West, both in terms of electric kitchen appliances and the production and consumption of convenience foods. Denmark was also slower to adopt many kinds of convenience food than the other Nordic countries, despite its high rates of female labour force participation and relative affluence. So there is no clear-cut relationship between the history of work, technological innovation and food preparation. Rather, more complex and locally-differentiated ways of introducing and incorporating convenience foods into regional and national markets prevailed. This is also true for different types of convenience food that are considered most typical in different national contexts (such as ready meals in the UK and frozen pizza in Norway).

Notes

1. Bruegel (2002, p. 126) uses the idea of 'normalization' to characterise this process, a term that we use in Chap. 4.
2. Lien (1997, p. 169) claims that Norway has the highest per capita sales of frozen pizza in the world.
3. These findings appear to contradict France's reputation as the home of *haute cuisine* (Trubek 2000; Ferguson 2006), suggesting that everyday cooking may be significantly different from fine dining in high-end restaurants.
4. Another MINTEL study reported that three quarters of UK adults had purchased a prepared meal in the last six months (Prepared Meals, May 2013).
5. In the UK, men and members of one-person households were reported to be the most likely to use ready meals, while women, 16–24 year olds, over-55s and members of two-person households were the least likely (MINTEL, Chilled and Frozen Ready Meals, May 2010).
6. Hecht (2009, p. 15) writes: 'I use this term to refer to the strategic practice of designing or using technology to constitute, embody, or enact political goals. Here I define technology broadly to include artefacts as well as non-physical, systematic methods of making or doing things'. Note that this notion of technopolitics fits well with our praxeological approach. The early promotion of convenience food and its supporting material infrastructure, as well as the competences and skills needed, may just as well be described as technopolitics.
7. Susan Reid (2002) has described the kitchen as a crucial site in the de-Stalinization of consumer taste in the Soviet Union following the encounter between Khrushchev and Nixon in 1959.
8. For more on the significance of 'cozy Friday' in Sweden, see: https://sweden.se/culture-traditions/cosy-friday/ (accessed 20 December 2017).

3

Convenience Food as a Contested Category

Introduction

Convenience foods can take many different forms in particular social and cultural contexts. Chapter 2 showed that the market for convenience food has developed both quantitatively, through increased market share of the food sector, and qualitatively, through product differentiation in terms of supply. An illustration of this from Denmark can be made by comparing what kinds of convenience foods were available 30 years ago and now, even though the Danish convenience food market is a relative late-comer compared to some of the other European food sectors, and Danes tend to be the least likely to buy ready-made dishes (Euromonitor International 2016a). In the 1980s, for example, ready meals in Danish supermarkets consisted largely of tinned casseroles and a few frozen dishes such as pizzas and spring rolls, together with 'the Saturday chicken', discussed in Chap. 2. Today, the freezers and fridges of Danish supermarkets contain a large variety of ready-made dishes from different ethnic or national cuisines, and dishes belonging to various dietary regimes such as organic, vegetarian and paleo.

P. Jackson et al., *Reframing Convenience Food*,
https://doi.org/10.1007/978-3-319-78151-8_3

The development of the European market for convenience food described in Chap. 2 could point to convenience food having become institutionalized as an ordinary and accepted part of consumers' everyday lives. However, existing empirical studies of consumers' use of convenience food suggests a less simple picture, where convenience food has been subject to problematization as well as institutionalization. At the same time, convenience food remains a difficult category for researchers to define adequately.

This chapter begins by reviewing the literature on convenience food. This is done in two ways: first, with regards to the socio-cultural status of convenience food among consumers, and second, with regards to the definitional status of convenience food among academic researchers. In the later part of the chapter, building on the insights of our practice-theoretical perspective on food in everyday life, we suggest a new conceptualization of convenience food. We illustrate this reframing of convenience food by comparing the four cases of baby food, ready meals, canteen food and meal-box schemes from the FOCAS project.

Convenience Food as a Socio-culturally Contested Category

When looking into how consumers tend to treat convenience food in their ordinary daily lives, an ambivalent relation seems to be being expressed. On the one hand, convenience food products and procedures such as ready meals, semi-processed foods (such as sauce powders and cake mixtures) and take-away food seem to have become a relatively routinized and expected part of providing meals in consumers' everyday lives (Jabs and Devine 2006; Olsen et al. 2009), indicating that convenience food may have become part of what is considered a 'normal' way of practicing food (see also Chap. 4 on normalization). On the other hand, the uses of convenience food products and procedures in consumers' food practices seem to be accompanied by normative negotiations and legitimations (Bugge and Almås 2006; Carrigan et al. 2006), suggesting that convenience food is not necessarily seen as a socially acceptable way of providing meals.

We begin with those studies, across a range of Western national contexts, that illustrate the tendency for convenience food to have become *routinely used and expected.* All of these studies show either quantitatively or qualitatively that consumers regularly use and expect others to use convenience food as part of their food practices. Some studies indicate that the sheer amount and regularity of convenience food use in households is sufficiently large to regard convenience food as part of the ordinary range of food choice. This is the case, for example, in two US studies. An overview article on the implications of time-scarcity on cooking and eating practices by Jabs and Devine (2006) shows that there has been a decrease in food preparation at home and a parallel increase in fast-food consumption, with around one third of American parents stating that they regularly eat take-out meals. A qualitative study shows how attempts to integrate work and family life have consequences for parents' ways of managing food provisioning, cooking and eating. One of the food choice coping strategies is to use convenience food (Blake et al. 2010).

Similar indications can be seen from studies that try to map a number of consumer segments who use convenience food in different ways. In a quantitative study in Belgium, for example, the researchers identified five different segments among Flemish consumers according to their relation to ready meals: energetic experimenters, harmonious enjoyers, normative carers, conscious experts, and rationalists (Geeroms et al. 2008). A quantitative study in Norway, also about consumers' use of ready meals, delimited three segments: convenient consumers, ambivalent consumers, and dissatisfied consumer (Olsen et al. 2009). Regardless of the number of segments, both studies indicate that large numbers of consumers use convenience food and are able to relate to the category of ready meals (albeit in different ways).

Several qualitative studies and review articles describe different ways in which consumers routinely expect varieties of convenience food to be part of their meal provisioning. An American review article argues that convenience food use should be seen in the context of the division of household labour and the reproduction of everyday life, where convenience food can be seen as a way for women to handle food-work as part of their busy working lives (Szabo 2011). A qualitative study of food routines among mothers in the UK describes how women understand convenience items and procedures as lending order, control and autonomy in food-work, and that convenience food can be seen as 'a good compromise' in providing meals (Carrigan and Szmigin 2006). Parallel results come from a qualitative study of food practices among women from New Zealand, where convenience food was described as a device that helped women overcome some of the constraints to carrying out their preferred food practices, such as time-pressure, the unpredictability of everyday life, and a lack of cooking skills (Bava et al. 2008). However, this kind of routine expectation of convenience food as part of meal provisioning does not only seem to take place in busy family lives, but also among other kinds of consumers and in different kinds of households. For example, a qualitative study of young (20–25 year-old) Danish consumers' food routines showed that convenience food is experienced and understood as something everybody does among young people (Halkier 2017a), something also shown in a Dutch study of adolescents (Wahlen et al. 2016).

Another way of looking at how convenience food can be seen as expected in 'normal' food conduct is to analyse the similarities and differences in meals that involve convenience food and those which do not. For example, a survey-based study among Swedish consumers in the second largest city concerning the use of ready meals concluded that meal situations with ready meals did not differ very much from other meal situations—except from the greater tendency for ready meals to be eaten alone (Ahlgren et al. 2004). Another qualitatively-based Swedish study described how going to fast-food outlets could actually be seen as supporting the reproduction of traditional family life rather than threatening to undermine it (Brembeck 2005).

Finally, some previous studies highlight how particular types of food, associated with convenience, can be expected as part of people's food practices, due to their materiality and taste. In a French quantitative study of the use of different kinds of vegetables, for example, consumers tended to use particular frozen vegetables as willingly as fresh ones (Plessz and Gojard 2015). In an American cross-generational qualitative study of how consumers constructed their family identity and cooking practices through home-made food, younger generations tended to express to a higher degree how convenience food could be more tasty than home-made food, depending on the type of food (Moisio et al. 2004).

However, patterns of understanding and handling convenience food in some of these and a range of other studies have pointed towards the moral questioning and justification associated with convenience food, suggesting that convenience food use is perhaps not necessarily socially accepted as part of ordinary food conduct. This tendency was noted in Alan Warde's oft-cited comment that 'the idea of convenience food is tinged with moral disapprobation' (1999, p. 518). In all the previous studies reported below, convenience food was, in one way or another, linked by consumers with less acceptable characteristics, such as being low quality industrial products, less healthy, less sustainable and, most of all, seen as an inappropriate way to provide food and meals for family members.

The tendency for consumers to understand convenience food as not good enough can be found in a number of the previous studies. In a quantitative study, for example, Swiss consumers were reported as having negative attitudes towards ready meals (Brunner et al. 2010). A Swedish survey-based study likewise concluded that consumers did not necessarily see ready meals as of sufficient quality to serve for dinner (Prim et al. 2007). A Finnish study of consumers' relations with local food described their constructions of convenience food as inauthentic, in contrast with 'do-it-yourself' food such as locally provisioned raw ingredients including mushrooms, berries and fish (Autio et al. 2013). The few studies which look at several generations of consumers' understanding of convenience food or consumers' construction of people from other generations in relation to convenience food have tended to conclude that older generations worry about the quality of food provided by younger generations, including the amount of convenience food consumed. The American study of homemade food and family identity across three generations referred to above (Moisio et al. 2004) described the elderly generation as not wanting to include convenience products in cooking for their families, and in a Canadian study on traditional food knowledge, older women constructed younger generations as not being sufficiently skilled to cook because of their convenience food use (Braun and Beckie 2014). However, even among younger generations, research has also reported understandings of convenience food as not good enough. In the Danish qualitative study mentioned above

(Halkier 2017a), young consumers constructed some forms of convenience food (such as frozen ready meals) as less acceptable than others.

The existing literature also includes a number of studies which describe normative negotiations and justifications being made by consumers in relation to their use of convenience food. A qualitative study of domestic dinner provisioning among Norwegian women with children described the construction of so called 'proper dinners' being contrasted with 'poor dinners', based on ready-made products, and the mothers tended to have to legitimize dinner situations where they used convenience food (Bugge and Almås 2006). In a parallel UK study of mothers' food practices, convenience food was also constructed as 'not proper food', while at the same time being justified as food that could be necessary when, for example, the complex scheduling of everyday life activities demanded it, where conflicts over food tastes in the family occurred, or when mothers felt that their food work was not appreciated (Carrigan et al. 2006). Another qualitative study of gender and UK kitchen practices described among other things that everyday cooks made what they called 'unapologetic apologies' for their use of convenience food for which they imagined they would be morally judged (Meah and Jackson 2013—see also Chap. 7 on the moralization of convenience food). In a qualitative study of cooking in Canada, one of the main results was that cooking home-made meals gave the participants an experience of control over the family food intake, enabling them to keep family members away from the negative associations of certain types of food, such as fast food, convenience food and processed foods (Simmons and Chapman 2012). A Finnish qualitative study of online conversations about convenience food described how the participants negotiated the normative ambivalences of 'appropriate methods' in food provisioning and preparation (Närvänen et al. 2013).

Summing up, existing research on convenience food-use displays two main tendencies. First, convenience food seems to have become routinely used and socially expected among consumers across different cultural contexts in the Western world. Secondly, however, this does not necessarily mean that convenience food is a socially and normatively accepted part of ordinary food conduct among consumers. Convenience food is still associated with less good food quality and with not living up to normative ideals about cooking from scratch.

Convenience Food as an Analytical Category

The previous section argued that convenience food is a socio-culturally contested category. There is no clear definition of convenience food in academia either (cf. Bava et al. 2008; Jackson and Viehoff 2016; Scholliers 2015). Some researchers gloss over this problem by working with an almost implicit taken-for-granted understanding of convenience food as being equal to ready meals (Botonaki and Mattas 2010; Costa et al. 2007; Geeroms et al. 2008; van der Horst et al. 2010; Mahon et al. 2006; Olsen et al. 2010; Reed et al. 2000, 2003). Others attempt to define convenience food with nominal definitions, consisting of particular food items (but broader than just ready meals) that are listed and organized as, for example, convenience, semi-convenience and non-convenience foods (Daniels and Glorieux 2015). A well-known definition of convenience food comes from the article by Costa et al. (2001) where convenience is defined at different stages, ranging from 'ready-to-eat', 'ready-to-heat', 'ready-to-end-cook' and 'ready-to-cook'. Such attempts have been criticised for broadening the category too much, with nearly all kinds of food and meal-types included such that the category of convenience food cannot discriminate analytically. In a recent special issue of the journal *Appetite* on convenience food, Scholliers (2015) draws upon this critique in order to underline how both academic and socio-cultural definitions of convenience food will probably always be dependent upon historical and socio-cultural contexts.

One strategy for taking into account the dependency upon social and cultural contexts in the definition of convenience food has been to use the mundane definitions from food consumers themselves who participate in research projects on convenience food. This is the case, for example, in a study of UK convenience food consumers by Buckley et al. (2007). The same strategy is used by Carrigan et al. (2006) in the study referred to above about UK mothers and convenience food. To use mundane definitions as a starting point for academic definitions is a familiar strategy in qualitative research. But it tends to work better in English-speaking countries as 'convenience food' does not translate easily into other languages. In Danish, for example, the English term translates as either *sammensat fødevare* (compound foodstuff) or *convenience mad* (convenience food), while other relevant Danish culinary categories include *færdigmad* (ready-made food), *hurtigmad* (fast food), *nem mad* (easy food), *halvfabrikata* (processed food) and *tage-med-mad* ('to go' food). In Swedish, 'convenience food' translates as *lättlagad mat, snabbmat, bekvämmat* or *färdigmat* (respectively, easily-made food, fast food, comfortable food or ready-made food), while in German, the most common comparable terms include *Fertiggericht* (ready/instant meal), *Fertigessen* (ready food) and *Fertigfras* (ready-grub). The English term 'convenience food' is often used as a technical term by people in the catering industry and food retail sectors across Europe to circumvent these translational issues.

A further definitional challenge consists in the relationship between convenience food and another food category, namely 'home-made' food or food cooked 'from scratch'. Most often, convenience food is defined in contrast to home-made food or meals cooked from scratch (Autio et al. 2013). However, the research literature does not provide a clear definition of 'home-made' or 'cooked from scratch' either (Halkier 2009; Moisio et al. 2004; Short 2006; Slocum et al. 2011). To establish under which circumstances a food item crosses the line from raw ingredient to convenience product or when a cooking procedure crosses the line from cooking to re-heating, for example, is just as dependent upon historical and socio-cultural contexts.

In the UK, for example, the Food Standards Agency provided official guidance on the commercial use of the term 'home-made'. Acknowledging dictionary definitions of the term as referring to food that is made or prepared in the home; of domestic manufacture; made at home using traditional methods rather than by a manufacturer; made by oneself; or crudely and simply made, the FSA argued that consumers understand the term to mean food prepared in a domestic kitchen rather than in a factory or manufacturer's kitchen. The Agency also recommended that factory-made foods should avoid misleading visual representations of farm-house or small-scale domestic kitchens. Finally, they advised that the designation 'home-made' in catering establishments should be restricted to the preparation of recipes on the premises using primary ingredients, in a way that reflects a typical domestic situation (FSA 2008, pp. 81–84).

In light of these definitional difficulties, some researchers have encouraged a discussion about how to reframe convenience food as an academic category. An early example comes from an article about eating patterns among UK and Australian students (Marshall and Bell 2003) where the authors argue for working with a broader definition of meals than is often used by researchers, whereby home-made food and convenience food are seen not as two separate categories but part of a continuum of eating events. Another more recent call for reframing the category of convenience food is based on an argument for seeing convenience food as part of the overall division of labour in food work, not just inside households and among families but also in society as a whole. Szabo (2011) argues that convenience food relates to the de-routinization and time-pressuring of gendered household work, and that it shifts the allocation of tasks within and beyond the home, including the shift of food-work to places of production and commercial provisioning of food, such as farms, factories, stores, restaurants and canteens (where such work is often undertaken by working-class women from racialized minority groups).

As part of the current study, Jackson and Viehoff (2016) argued that convenience food is an example of a 'chaotic conception', which divides items that are logically indivisible while lumping together things that are logically distinct, meaning that such terms should not be used for explanatory purposes. Reviewing more than 100 papers on convenience food, they identify a number of key themes including time-saving and time-shifting, family and motherhood, health and nutrition, and cooking skills. They conclude, like Marshall and Bell (2003), that convenience food and home-made food are not separate categories and that the category of convenience food should be embedded in an understanding of everyday food practices and overlapping contextual arrangements.

A further argument for reframing convenience food as an academic category can be made on the basis of a particular empirical type of pattern in the food and everyday life literature. When people are providing food, when they are cooking and eating, they often tend to combine different kinds of food, provisioning strategies, cooking procedures and eating activities. A family can have an ordinary dinner consisting, for example, in the heating of a ready-made quiche from the freezer in the supermarket, a salad made with home-grown lettuce, herbs and flowers from their garden, and a soup bought in a work canteen on the way home. The following publications all make this point about the hybridity of food provisioning, cooking and eating on the basis of empirical research.

Examples from the UK include the 'mothers of invention', described in the qualitative study by Carrigan and Szmigin (2006), who juggle and piece together meals through artful combinations of fresh and convenience ingredients. A similar picture emerges from a Danish qualitative study, where the author showed that female lifestyle-magazine readers where using, adapting and negotiating cooking from scratch and convenience food procedures. One of the women expressed it this way: 'So a homemade soup is completely made from scratch, but preferably in an easy way, not like having to boil a chicken for three hours [laughs]' (Halkier 2009, p. 372). In a qualitative study of weaning practices in Sweden (as part of the current project), the authors show that what was served to babies was most often a mix of ready-made and home-made food (Brembeck and Fuentes 2017). Based on quite different data, namely quantitative time-use survey data, Mandemakers and Roeters (2015) suggest that Dutch consumers spend significantly less time on cooking and eating during weekdays but more at weekends, indicating that meal provisioning and eating are a mix between less and more time-consuming procedures over the course of a week. Another example from a qualitative study in the UK (Hand and Shove 2007) illustrates some of the details in hybridised ways of providing meals. The study focuses on the uses of freezer technology and one of the results is that consumers tend to use their freezers to combine home-made food items with convenience food in their attempts at organizing and timing meals in everyday life. In the same direction, a UK study on cooking (Short 2006) suggests that the academic definition of cooking skills should be broadened to cover planning, organizing and assembling skills, procedures that are often also connected with convenience food solutions.

To sum up: just as convenience food is a socio-culturally contested category, there is no clear definitional conceptualization of convenience food in the research world either. The definitional debate covers several issues. First, definitions of convenience food tend to depend upon historical and socio-cultural contexts. Second, the actual term 'convenience' travels badly across languages, raising challenging translational issues. Third, convenience food as an academic category is often positioned as the opposite of home-made food, but home-made food is an equally unclear and debated category. Finally, research across contexts using different kinds of data suggests that food practitioners combine different food types, provisioning strategies, cooking procedures and eating patterns, thereby blurring categories and creating further definitional issues.

From 'Convenience Food' to 'Conveniencization'?

Following attempts to reframe the term convenience food (Jackson and Viehoff 2016; Halkier 2017a), we propose to embed convenience food in a theoretical understanding that places convenience food as part of food practices and processes, using a 'theories of practice' perspective (as outlined in Chap. 1). Here, we summarize some of the relevant main assumptions of the approach, before introducing the concept of 'conveniencization'.

First, a practice-theoretical approach focuses on the organization and accomplishment of mundane practices. Second, the unit of analysis is practices and ways of practicing. Thirdly, every context consists of a multiplicity of different intersecting practices. Fourthly, material arrangements are important parts of practices and practicing. And finally, practicing takes place on the basis of notions of socially expectable and acceptable conduct.

Rather than seeing convenience food as a particular type of foodstuff, we introduce the notion of 'convenient' food provisioning, cooking, eating and wasting. Convenient here means easily provided food, readily made into a meal, easily eaten and/or easily dealt with as waste. One of the advantages of the term 'convenient' over the idea of 'convenience' is that it travels much better across languages. 'Convenient' has clear proxies in the three non-English speaking contexts of the FOCAS project. In Danish, convenient equals 'easy food' (*nem mad*), in German *Fertiggericht* and, in Swedish, *bekväm mat*. Another advantage of 'convenient' over 'convenience' is that 'convenient' is a process term, and thus more dynamic to apply in different socio-cultural contexts and across a variety of cases. An example from a Danish study with young food practitioners (Halkier 2017a) showed that traditional 'convenience' food was only one out of three different 'convenient' ways in which young people acquired food, cooked and ate.

Understanding convenience food in terms of convenient food provisioning, cooking, eating and wasting enables us to reframe convenience food in terms of the process of *conveniencization* by which we refer to the way certain kinds of food, in certain circumstances and in particular combinations, come to be regarded as convenient.[1] We understand conveniencization as covering the different degrees of convenient-ness of food and food-related practices as well as the different ways in which food practices can become convenient. Thus, conveniencization is also a process term like 'convenient' which it builds upon. However, conveniencization adds directionality in terms of the dynamics of food-practicing processes: how much ease in food provisioning, cooking eating and wasting? Conveniencization also adds a capacity to distinguish varieties of easing food practicing in the hybridity of food practices: what kinds of ease in food provisioning, cooking, eating and wasting?

The concept of conveniencization both replaces the analytically problematic term 'convenience' and embeds the notion of convenient in a process-based perspective on mundane practicing, able to investigate both directionality and variation. Such a conceptual replacement makes for a greater definitional adaptability to the hybrid character of food provisioning, cooking and eating, which was argued in the former section. If food practitioners combine different kinds of convenient food elements and procedures with different kinds of raw ingredients and cooking from scratch procedures, then researchers should be able to understand such combinations more thoroughly. Furthermore, a shift from 'convenience' to 'conveniencization' also enables better definitional adaptability to the apparently growing differentiation of 'convenience'. If 'convenience' increasingly also covers healthier, organic, paleo and related food practices and if convenient ways of providing meals increasingly also cover food bags and meal-boxes, for example, then it is necessary to work with an understanding which allows for greater variation in the phenomenon.

As a way of summing up, the relevant main assumptions in a practice-theoretical approach can now be specified for conveniencization. *First*, conveniencization processes take place as part of the organization and accomplishment of mundane food practices. Such food practices both form a socially recognizable entity and are being continually performed as food provisioning, cooking, eating and wasting activities. *Second*, conveniencization denotes the degrees and kinds of ease with which food provisioning, cooking, eating and wasting activities are being practiced. *Third*, degrees and variations of conveniencization in different elements of food practices are related to the ways in which food practices intersect with other mundane practices, such as working, travelling, parenting and socializing. *Fourth*, material arrangements around food provisioning, cooking, eating and wasting help form degrees and types of convenient-ness in food practicing. Examples of such material arrangements could be as different as freezers, gardens, shops, canteens and internet access. *Finally*, the degrees and kinds of convenient-ness experienced by food practitioners are part of the process of socially adapting to, experimenting with and negotiating expectable and acceptable food conduct. Thus, different degrees and kinds of convenient provisioning, cooking, eating and wasting configure differently with food's normality and normativity.

Comparing Conveniencization

In this section, we use the conveniencization perspective to provide an initial comparison of the degrees of convenient-ness in food practicing and variations in how food practicing becomes convenient within and between our four cases (baby food, ready meals, canteen food and meal-box schemes). This comparison is preliminary and much more detailed comparisons are provided in the empirical chapters that follow, using the 'constant comparative method' of qualitative analysis (Hammersley and Atkinson 1995, pp. 232–236; Silverman 2006, pp. 295–297). This preliminary comparison, however, is not a traditional social scientific one, comparing either variables as in quantitative comparison or whole cases as in qualitative comparison (cf. Krause 2016; Ragin 1987). Our purpose is not to seek causal explanations or simplifying parallels, especially not at the level of 'national culture'. The book neither contains case-bound chapters, such as 'the baby food case' or 'the canteen case', nor nation-bound chapters on 'the British case' or 'the Danish case'.

Rather, our comparisons are qualitative socio-cultural comparisons, following current thinking about making comparisons at meso-level and in more analytically sensitized ways (cf. Deville et al. 2016; Lamont and Thévenot 2000). The focus has been on keeping description, concept formation and interpretation explicit, shared, sensitized and relatively 'straight' in our team of researchers (cf. Krause 2016). For example, we have thought through the different aspects of conveniencization by applying the same categories to the quite different kinds of data within and across the case studies. This has helped produce an initial frame for more detailed empirically-informed comparison of specific relevant social dynamics such as normalization and moralization (explored in detail in Chaps. 4 and 7).

Discussing degrees of convenient-ness across our four cases is clarified by examining where and when food provisioning, cooking, eating and wasting seem to be convenient (see Fig. 3.1 which proposes a typology of ready-to-heat, -eat or -cook, building on the work of Costa et al. 2001).

Comparing *what* is considered convenient in the different ways of providing, cooking and eating different kinds of convenience food shows that ready meals, baby food and canteen food share the characteristic of requiring little food work, whereas meal-box schemes are not ready to eat or heat, but need to be cooked first. This might lead us to conclude that meal-box schemes are the food activity with the least degree of convenient-ness. However, when the dimension of *where* is included in the comparison, both canteen food and meal-boxes can be seen as exhibiting a lesser degree of convenient-ness, considering that both of them are primarily used in one location, whereas both ready meals and baby food tend to be able to travel across several locations.

If we include *when* a food activity is convenient in the comparison, it becomes more difficult to say something relatively clear about parallels and differences in degrees of convenient-ness between the four food activities, because 'when' refers to a wider range of different procedures in food work, than the 'what' and the 'where'. Ready meals, baby food and canteen food all share the characteristic of being convenient when other food provisioning procedures are not do-able. For example, parents described how commercial baby food allowed them to feed the baby while out walking or in the car, but also at home when the baby needs

	Baby food: ready to heat or eat	Ready meals: ready to heat or eat	Canteen food: ready to eat	Meal box-schemes: ready to cook
Where convenient	At home, on the road, visiting	At home, at work	At work, sometimes at home	At home
When convenient	Planning; other procedures in provisioning not do-able; make timing flexible for other activities; handling potentially constant provisioning	Planning; other procedures in provisioning not do-able; make timing flexible for other activities	Planning; expected routine with others; other procedures in provisioning not do-able	Planning; avoiding the need to shop in physical shops; make timing flexible for other activities; preparing dinner

Fig. 3.1 Conveniencization compared

immediate attention and care, making cooking more or less impossible. Meanwhile, participants in the UK and Germany described how ready meals gave them flexibility in the timing of meals for different family members, allowing them to keep a 'stock' of food to cater for unexpected visitors and providing scope to shop and eat at different times, using freezer and microwave technologies to time-shift as well as time-save (see Chap. 5 on the temporalities of convenience food). While chilled ready meals are usually bought for consumption within a couple of days, they can be frozen and stored for several weeks, like foods that are already frozen when purchased.

On the other hand, meal box-schemes, baby food and ready meals have the characteristic in common of being convenient in helping to make timing flexible for other activities in everyday life. Parents described how using a meal-box scheme enabled them to have time for helping kids with home-work, doing other types of housework, working longer hours or simply relaxing, because they did not have to spend time physically shopping for food for the family meals. Similarly, UK participants justified the use of ready meals in terms of reducing the amount of time spent cooking and cleaning up so that more time could be devoted to other (more valued) domestic responsibilities, sometimes described in terms of spending 'quality time' with other family members. Ready meals were also used to provide food for other family members when those with primary responsibility for cooking were not available, as discussed by Meah and Jackson (2017) in terms of 'convenience-as-care'. However, one temporally-related dynamic, which all the four types of food provisioning activities have in common in terms of when the activities are (at least potentially) regarded as convenient, is when planning the processes of provisioning, cooking and eating in everyday life. This does not mean that food practitioners always plan ahead, but if they do, then baby food, ready meals, canteen food and meal-box schemes all seem to have the potential for making planning more convenient.

Discussing the various ways in which food activities can become convenient is clarified by looking at how food provisioning, cooking, eating and wasting are regarded as convenient across our four cases. In order to make the comparison more detailed, we have divided the 'how' into three dimensions, which refer to the three overlapping aspects of consumption suggested by Warde (2010): acquisition, appropriation and appreciation (see Fig. 3.2).[2]

'Easy' is a word that is often used to describe something being convenient, but it is used in a number of slightly different ways by the participants in our four case studies, covering, for example, fast, simple, uncomplicated, well-known, and flexible. Regarding how food *acquisition* becomes convenient, ready meals, baby food and canteen food all involve much less food shopping, due to the provision of a whole meal in one act of purchase. Regarding ready meals and baby food, food can be bulk-shopped and stored in freezers, fridges and cupboards, whereas meal-box schemes can mean no shopping, except via an internet-connected device such as a PC, tablet or mobile phone, while baby food products assist parents in planning, shopping and storing the baby's meals. Parents describe commercial baby food as a vehicle for meal variation and securing appropriate nutrients and texture. The ability to store food and keep it at hand is also repeatedly described as a convenient quality of baby food. For example, Sara describes *klämmisar* [food pouches] as 'very neat to bring as emergency food' and Johanna describes baby food tins as 'very useful if you are off somewhere'.[3]

Convenient how	Baby food	Ready meals	Canteen food	Meal-boxes
Acquisition	Less shopping (for separate ingredients); more choice; less planning	Less shopping (for separate ingredients); more choice; less planning	No planning; less shopping; limited choice	Planning organized; less shopping in store; limited choice
Appropriation	Enhancing mobility; nearly no cooking; easy serving and eating; less wasting?	Nearly no cooking; easy serving and eating; less wasting?	No cooking; easy serving; built-in variation of dishes in material arrangements; social eating	Directed cooking; built-in variation of dishes in material arrangements; eating homemade meals; less wasting?
Appreciation	Relief from normative anxieties and reflections; enables the serving of homemade meals; socially expected	Enables time to be spent on other activities; enables different household members to eat differently; ambivalent acceptance	Enables time to be spent on other activities; socially expected; ambivalent acceptance	Relief from normative anxieties and reflections; enables the serving of home-made meals; normatively accepted

Fig. 3.2 The acquisition, appropriation and appreciation of convenience food

Looking at how *appropriation* becomes convenient, the comparison of the four cases suggests a kind of continuum, but also parallels between the cases. The continuum in how appropriation of food becomes convenient goes from canteen food involving no cooking, ready meals and baby food involving nearly no cooking, and meal-box schemes involving what might be called directed cooking, since the meal boxes arrive with step-by-step recipes and already-measured ingredients for each meal. On the other hand, canteen food and meal-box schemes involve a pre-selected (by others) variation of food and meal items in the material arrangements that ready meals and baby food do not exhibit to the same extent.

How eating is convenient again differs, since ready meals and baby food tend to be easily eaten. Many parents describe how their babies liked commercial baby food and many named specific products as their baby's favourite food. For example, Camilla, a mother of four, described Semper's 'Rice, chicken and vegetables' as her one-year-old daughter's favourite while Julia, a first-time mother, described her one-year-old son's preference for Hipp vegetable lasagna. For Camilla, Sara and other parents like them, commercial baby food is an easy way to acknowledge the babies' preferences and boost their appetite.

Eating canteen meals tends to be socially convenient, providing an opportunity to socialize with colleagues while eating. What adds to the social aspect of canteen food is the fact that coffee and lunch routines are often established practices which do not require time-consuming planning in advance; rather, members of these practice communities can built on the shared knowledge that the other members are going to be around in the canteen at the usual time. As a female judge in a courtroom canteen remarked: 'Actually, we always have lunch here, don't we? That is, when we have time, when we are not deterred by hearings. Lunch, in fact, is very important for (verbal) interchange'. Here, the interviewee does not only refer to social interaction among colleagues, but also to an informal exchange of information and easily accessible advice. By contrast, eating meal-box food tends to be normatively convenient in the sense that the food practitioners can claim to have eaten home-made meals, which is a normative standard that people try to live up to (Hertz and Halkier 2017).

Finally, when comparing how *appreciation* becomes convenient across the four cases, a different set of parallels and contrasts can be seen. This time it is baby food and meal-box schemes which have types of convenientness in common. Both food activities tend to be appreciated as convenient, due to their apparent ability to relieve food practitioners from normative anxieties and reflections about whether the food served is sufficiently healthy and environmentally sustainable. Commercial baby food was repeatedly described as a means for securing the nutritional needs of babies and opting for an eco-labeled or locally produced product was described as a way to see to the baby's needs and wider societal concerns at the same time. Meal-box-schemes were repeatedly mentioned as 'outsourcing' normative responsibilities for nutrition, health, culinary variation, sustainability and reduction of food waste.

Ready meals and canteen food tend to be interpreted in more ambivalent terms regarding how normatively acceptable the food is seen to be. So, for example, participants used self-deprecating humour and irony to 'excuse' their use of ready meals, describing the embarrassing moments when they were 'caught out' using convenience foods or feeling the need to apologize for taking culinary short-cuts such as the use of prepared pastry or ready-made sauces.

In canteen dining rooms, an individual's food intake is on public display: everybody can see what and how much a person is consuming. This frequently leads to justifying reactions regarding the kind of meal chosen, especially in relation to the calorie content, fat, salt and sugar in hearty meals. At the same time, dishes like currywurst or schnitzel with French fries are among the most popular canteen dishes in Germany. This means that, although the consumption of these meat-containing dishes, high in calories and salt, is a frequent practice, it requires an 'excuse', at least in terms of lip service. Participatory observation reveals, additionally, that canteen food in particular and diets in general are a common topic of lunch-time conversation. Together with shared ideals of body-size and the fact that the actual consumption habit is openly on display, this creates a form of peer pressure, influencing the everyday experience of having lunch in a canteen. On the other hand, ready meals, canteen food and baby food enable food practitioners to spend time on other activities and can be regarded as convenient in that sense.

Conclusion

This chapter has shown how convenience food is a contested category in academic usage (as an analytical term) and in everyday life (as used by consumers). It has demonstrated how convenience food has become *normal* in the sense of becoming a routine and socially expected part of everyday food consumption but is still regarded ambivalently in *normative* terms. We have proposed a distinction between convenience food, convenient food and conveniencization, consistent with our commitment to a 'theories of practice' approach. While the term 'convenience food' is conventionally applied to specific kinds of food, such as ready meals or frozen pizza, our research shows that many different kinds of food can be regarded as 'convenient' (whether or not they fall within the category of 'convenience food'). We have therefore introduced the idea of 'conveniencization' to refer to the process by which different kinds of food, on different occasions and in different circumstances, come to be regarded as convenient. This helps avoid the moralization of certain kinds of food, taking a less judgmental approach to the practices of preparing, eating and disposing of food. It also enables us to focus on the different points (in time and space) that 'convenience' can be enacted, whether at the planning and provisioning stages, during cooking and eating, or in relation to clearing up and disposing of food. All of these issues are developed in more detail in subsequent chapters.

This chapter has sought to provide greater clarity in our conceptualization of convenient and convenience food, reviewing previous studies and providing a justification for the kind of comparative work that we have undertaken. We have also provided a preliminary socio-cultural comparison within and across our four cases. This comparison focuses on when, where and how the four different kinds of foods and food-related activities associated with baby food, ready meals, canteen food and meal-box schemes *become convenient* in terms of acquisition, appropriation and appreciation.

Notes

1. We are grateful to Monica Truninger and Rafi Grosglik from the Sociology of Consumption research network of the European Sociological Association for proposing the term 'conveniencization' when commenting on an early presentation of our ideas at a conference in Porto in 2014.
2. In brief, acquisition is about how food and meals are procured; appropriation is about how food is used and integrated in everyday contexts; and appreciation is about the interpretation of food.
3. Details of all the participants are given in the Appendix.

4

The Normalization of Convenience Food

Introduction

Despite being framed as morally ambiguous, convenience foods are today commonly integrated into consumers' everyday routines and food practices. Although many consumers might aspire to ideals of 'cooking from scratch' and the use of 'fresh' ingredients, previous research indicates that convenience products make up a substantial portion of contemporary food provisioning and cooking (Warde 1999; Halkier 2016). Using the consumption of commercial baby food in Sweden as a case study, this chapter outlines how convenience food is normalized, addressing the circumstances and resources that make the morally ambiguous category of convenience food a normal part of everyday food practices.

Acknowledging the dual meaning of 'normal' as something that conforms to a type, standard or regular pattern (for example, food that a large part of the population consumes regularly) and as something that does not deviate from norms, rules or principles (for example, food that is considered appropriate to eat), we discuss how commercial baby food is made to be both a *common and expected* and *proper and accepted* part of infant weaning practices in Sweden (cf. Brembeck and Fuentes 2017).

P. Jackson et al., *Reframing Convenience Food*,
https://doi.org/10.1007/978-3-319-78151-8_4

Following previous work on practice theory and normalization (e.g. Hand et al. 2005; Shove and Southerton 2000; Shove and Pantzar 2005; Hand and Shove 2007), we emphasize the continuous and localized material 'making' of commercial baby food as 'normal'. We examine the history and contemporary use of commercial baby food in Sweden to explore how the development of this product category facilitates the 'conveniencization' of weaning practices. We aim to illustrate how commercial baby food offers ease of use in terms of provisioning, cooking and feeding and at the same time carries 'frames of action' that invite parents to provision, cook and feed in an appropriate manner, meeting the contemporary demands of good mothering. Hence, we argue that the normalization of commercial baby food involves continuous work where products are invested with moral capacities (Hawkins 2012) that are used in the enactment of 'good mothering'. As such, they can also be seen as 'political devices' prompting the articulation of various interests (ibid., p. 67) from baby health and child-centredness to sustainability, impacting on the habits of households and changing the prerequisites for being a parent and a consumer.

Theorizing the Normalization of Convenience Food

As a derivation from the term *normal* meaning 'constituting or conforming to a type or standard' or 'the regular, usual, typical or conventional', *normalization* refer to the action or process of making something standard, typical or conventional (Oxford English Dictionary).

In the social sciences, the concept of 'normalization' is generally used to denote processes through which products, ideas or practices become expected and accepted parts of everyday life (Hand and Shove 2007; Halkier 2016). The concept has been applied and developed in various scholarly works. Examples include Foucault's notion of normalization as a labour of division and an instrument of power where individuals are rewarded or punished for adhering to or deviating from a constructed ideal norm of conduct (Foucault 1995) and Normalization Process

Theory (May and Finch 2009), addressing work that actors do when a new way of, for example, acting or thinking is integrated into already established knowledge or practices.

Normalization also parallels other concepts such as *diffusion*, used in marketing and innovations studies to conceptualize the market spread of products (Arnould 1989) or *domestication*, used in studies of media and communication to conceptualize the acceptance or rejection of media technology (Berker et al. 2006).

More specifically, in consumption studies 'normalization' is understood as a socio-material process where products make alliances with other objects and ideas (Hand and Shove 2007, p. 81). Influenced by science and technology studies and practice theory, consumption studies often conceptualize 'normalization' as a historical process where products and their use are the result of wider socio-material networks that have evolved over time (Hand et al. 2005; Hand and Shove 2007; Halkier 2016; Shove and Pantzar 2005). Work on mundane consumer objects, such as freezers (Hand and Shove 2007) and walking sticks (Shove and Pantzar 2005) or consumer practices such as washing (Shove 2003) or showering (Hand et al. 2005), convincingly illustrate how 'normalization' (or 'becoming normal') is an integrative process involving materials, technological systems, meanings, understandings and doings that co-evolve over time.

For example, writing on the appropriation and normalization of freezers in the UK, Hand and Shove (2007) describe normalization as a process where consumer goods move from 'novelty to ordinary invisibility' by becoming part of *successful alliances of objects and ideas* (ibid., p. 81). Normalization is seen as the result of an integrative process where a certain object such as a freezer, is made to fit into wider arrangements, such as household cooking and eating practices and systems of provisioning, in a manner that supports but at the same time structures these practices and systems. Hand and Shove (2007) describe how the freezer is integrated with other objects, ideas and actions to form a nexus that has organizing and performative capacity. The freezer 'figures as an orchestrating node' where 'discursive, material and temporal aspects of daily life *condense* around the freezer' and various aspects of consumption and provisioning meet (ibid.).

From this perspective the normalization of convenience food would be considered a matter of products forming successful alliances with the organized nexus of elements that are involved in everyday eating practices (Warde 2016) or, as in the case of commercial baby food, the wider socio-material arrangements of weaning (Brembeck and Fuentes 2017).

Thus, normalization can be seen as a process of particular kinds of food conduct becoming taken for granted and accepted among food practitioners as part of the social organization of everyday life with food (Halkier 2016). Providing for food in a convenient manner can become 'normal', because food practitioners through interaction with others observe regular social patterns and come to expect this, while at the same time sharing an understanding of how accepted these patterns can be (Warde 2014, p. 293). The degree to which using convenience food has become 'normal' is related to how 'do-able' it seems to be (Halkier 2010, p. 36) or, in other words, a combination of the practicalities and the normative negotiations of convenience food.

Furthermore, one of Hand and Shove's (2007) main points is that normalization should not be seen as an end-result but as an on-going achievement that requires continual work. This work of normalization can be accounted as an historical process where the 'multiple histories' that make up the 'social biography' of specific domestic products is used to illustrate the formation of practice (Shove and Southerton 2000, p. 303; see also Hand et al. 2005). But it can also be described in terms of the on-going and continuous work (symbolic and practical) that is required for an object to maintain its position as 'normal' (Hand and Shove 2007). Hence, normalization can be conceptualized in historical terms, i.e. how mundane objects *become* normal, but also in present terms, as how objects are made to *remain* normal (cf. Hand and Shove 2007).

To understand how a product or artefact (in this case, baby food) can be made to fit into wider socio-material arrangements and re-organize these arrangements, we use the notion of 'scripts' and more specifically the concepts *prescription, de-inscription* and *re-inscription* (Akrich 1992; Akrich and Latour 1992).

The concept of 'scripts' has been used to underline the 'dialectical' and 'relational' achievement of normalization (Hand and Shove 2007, p. 82), defined as 'frames of action' that products carry and embody: 'the script of an artefact invites certain user behaviour while counteracting other behaviour' (Jelsma 2006). The concept originates in semiotic approaches to the analysis of settings, machines, bodies and texts and how various socio-material settings involve a form of path-building where 'one privileged trajectory is built out of an indefinite number of possibilities' (Akrich and Latour 1992, p. 260). The normalization of products here becomes an issue of 'path building' where material artefacts invite certain actions and hinder others through the formation of 'routes' in the socio-material landscape we live in, thereby influencing or 'translating' the actions of people (Jelsma 2006, p. 223).

Correspondingly, *pre-scription* refers to the prescriptive force of artefacts (Jelsma 2003, p. 106), i.e. they invite some actions while hindering others. This is linked to the design or setup of the artefact or device. Prescriptive force requires '*inscriptions* to be made by the designer in the hardware and the software of the artefact' (Jelsma 2003, p. 107, emphasis in original). Akrich and Latour emphasize the moral dimension of artefacts stating that prescription entails what an artefact 'allows or forbids from the actors—human and nonhuman—that it anticipates; it is the morality of a setting both negative (what it prescribes) and positive (what it permits)' (Akrich and Latour 1992, p. 261). Products are designed with an intended use and these intentions shape their use. For example, packaging can be understood as inscribed with various meanings and instructions for use but also anticipates certain competences (Wagner 2013). The pre-scriptive force of an artefact involves its material and symbolic design. Packaging offers a material arrangement that invites action and delivers competence. But it also offers a surface, a 'screen' that can be used for pre-scription (Cochoy 2004). Logos, colours, images and text define the product and give instructions, thereby shaping the use and consumption of the product.

De-inscription, in turn, is defined as the 'reaction' to the pre-scription of an artefact. Users of artefacts are not seen as passive but rather as active co-agents able to challenge intentions of designers and artefact prescriptions. Scripts have an impact on practice as material objects afford action in accordance with their symbolic and material design but as they are used, objects are also de-inscribed by their users who might contest and challenge their intended use.

It is also important to underline that pre-scription and de-inscription are coinciding rather than separate acts. As producers give artefacts a material and symbolic design they inscribe them with pre-scriptive forces. These pre-scriptive forces impact on consumers' use of products but they do not necessarily control them. Akrich and Latour (1992) write of the possibility of a gap between pre-scription and de-inscription that 'defines the presence or absence of crisis' (ibid., p. 261). Things might 'run smoothly' with no apparent distinction between pre-scription and de-inscription. Or there might be a gap signalling some form of conflict or crisis. As pre-scription and de-inscription take place in a setting, a socio-material arrangement that changes over time, the circumstances of these acts also move. This requires continuous work or movement that can be referred to as *re-inscription*, a 'feedback mechanism' (Akrich and Latour 1992, p. 262) that redistributes mutable elements. Inscription and re-inscription are the same thing but re-inscription brings movement and transformation to the fore. It is a 'sophistication of the setting' that handles a crisis and narrows or even closes the gap (ibid., p. 262).

For an artefact to be normalized, i.e. taken for granted and seen as appropriate, the gap between pre-scription and de-inscription must be managed in a manner that makes the artefact fit with the other elements of a practice. In order to become part of a 'successful alliance of objects and ideas' an artefact has to be used in a manner that compiles wider socio-material arrangements and therefore invites corresponding actions.

The sections that follow use the conceptual framework outlined above to illustrate how commercial baby food was and is 'made' into a normal part of everyday weaning. First, we offer a short historical account of the way commercial baby food was introduced to the market and how its close fit with wider socio-technical arrangements of weaning practice made it normal. We then turn to present-day practice and the work required to make commercial baby food remain a normal part of weaning by analysing specific and local performances of this practice.

Studying the Normalization of Commercial Baby Food

Commercial baby food is, in Sweden as in many other countries, commonly used by parents when introducing solids (cf. Gritsai 2001; Carrigan and Szmigin 2006; Afflerback et al. 2013; Atkinson 2014; Bentley 2014). Products are easily available in retail outlets, offering infant formula, ready-made porridge, packaged dinners and ready-made fruit smoothies. Many restaurants and cafés supply microwave ovens or other heating possibilities as well as high chairs for babies, and some even offer commercial baby food in tins or jars.

The Swedish market for baby food is described as mature but profitable (Euromonitor International 2016b), and although products have been contested at certain times, for example when there were reports of heightened levels of lead, arsenic and cadmium in rice-based baby porridge (Radio Sweden 2013), commercial baby food remains a common part of weaning practices among a large part of the Swedish population.

Public health authorities in Sweden also approve the use of commercial baby food. The Swedish National Food Agency and health-care officials describe commercial baby food and home-made food as equally appropriate options:

> You do not have to choose between cooking yourself or buying commercial baby food, both options have their advantages. It is easy to buy readymade food and the commercial baby food is of high quality and satisfactory nutritive value. If you want to make home cooked baby food neither complex recipes nor star-chef cooking skills are needed to make tasty and healthy baby food. (Trädgårdh Tornhill 2012, p. 1)

Sometimes, parents are even advised to feed their babies commercial rather than homemade porridge since it is enriched with various nutrients (Livsmedelsverket 2011, p. 19). Another illustrative example of the National Food Agency's view on convenience food is found in an NFA guide for child health-care personnel where ready meals and processed food are described as viable options to make sure children receive a healthy and varied diet. The NFA also advise health-care personnel to encourage parents to serve their children fish in order to secure a necessary intake of nutrients:

> In order to secure the child's intake of vital nutrients parents should be encouraged to teach their child to like fish. Usually there are one or two fish dishes that the child could appreciate. Some parents find it hard to prepare and cook fish and could therefore be in need of information on how quick and easy it can be. There are also many good semi-finished products and ready meals available. (Livsmedelsverket 2015, p. 27)[1]

As in the case of baby-porridge, convenience products and ready meals are framed as viable ways for parents to offer their children a varied and nutritious diet.

The common use of commercial baby food does not, however, mean that all consumers use these products to the same extent or in the same way. Besides variations among consumer groups, products are also used in various ways and for different reasons. Approaching the normalization of a product as encompassing variety and difference in use is more fruitful than looking for stringent or consistent use. Following Hand and Shove (2007) who describe the use of freezers as linked to 'imagined or real accomplishments of practice that are in turn bound up with prevailing discourses of home, food and "normal" family life' (ibid., p. 83), we approach the use of commercial baby food as intertwined with weaning practice and values and ideals on food, care and 'proper mothering'. Commercial baby food also offers a set of qualities and values linked to contemporary mothering ideals (Fuentes and Brembeck 2017). Moreover, these products are involved in parents' handling of the temporal orders of everyday baby care (Brembeck and Fuentes 2017) as well as the socially expected conduct of family life.

The analysis presented in this chapter draws from material gathered for a project on the use of convenience baby food in Sweden. The Swedish material involves interviews, consumer diaries, and participatory observation at a baby café in a small Swedish town carried out during the autumn of 2014, print material on weaning distributed by public health care officials, and web-based marketing material from four baby food companies (Semper,

Hipp, Ella's Kitchen, and Lovemade) that we have presented in detail elsewhere (Fuentes and Brembeck 2017; Brembeck and Fuentes 2017).

The interview study was conducted at a 'baby café' at a local health clinic in the small town of Falköping in western Sweden. The fieldwork included observations every second Thursday during the autumn of 2014, formal interviews with 19 parents and small-talk with staff and parents during feeding time (see the Appendix for further details). We also asked our participants if they would agree to answer additional questions by e-mail that we would send in packages until the child reached one year of age.

The project took an interest in the 'stocks of knowledge' that consumers use in their everyday dietary practices including published advice from health-care professionals and commercial baby-food marketing (see also Fuentes and Brembeck 2017). Our analysis also draws some comparisons with the project's other case studies on food-box-schemes in Denmark and ready-meals in the UK and Germany.

A Brief History of Commercial Baby Food

The introduction of mass-produced baby food in the form of canned and jarred vegetables and fruit dates back to the US in the 1920s (see, for example, Apple 1987; Bentley 2014; Levenstein 1988; Smith 2004). This was not, however, the first example of commercially produced 'convenience food' for babies. Already in the mid-nineteenth century, infant formula powders to be mixed with milk or water were invented as substitutes for mothers' milk by German chemist Justus von Liebig and Swiss pharmacist Henri Nestlé. The development was spurred by recent advances in chemistry, demographic changes, and demand for female workers in factories, as secretaries, domestic help and shop workers. Soon, doctors were proclaiming these products to be superior to the milk of wet nurses and, by the late 1880s, several brands were on the market (Apple 1987; Bentley 2014; Levenstein 1988). The most popular of these powders was Mellin's Food developed in England and manufactured in Boston, which was portable and easy to prepare by dissolving a certain quantity in hot water and then adding milk. One of the reasons for the success of this brand was the company's skill in profiting from new marketing techniques. Ads appeared in newspapers and free handbooks of infant care were distributed to readers of middle-class magazines. By the 1880s a number of sources spread the view that artificial feeding was both scientific and modern (cf. Levenstein 1988). This is an example of the new connection between consumer goods and modern mothers, as women were invited to express good motherhood via consumption. The 'consuming-mother' was born (see Cook 1995; Bentley 2014).

Also, even in a small, late-industrialized country like Sweden, there were already ads for Mellin's Food in the beginning of the 1900s. An advertisement in a ladies' magazine from 1905 argues that:

> Mellin's Food is an easily digestible nutritious food for children as well as for elderly people. Also recommended by doctors because of its property of unerringly increasing breast milk where this is insufficient. Can be purchased at any pharmacy. Samples and a brochure about feeding children free from Mellin's Food Depot, Malmo. (quoted in Brembeck et al. 2001, p. 36)

The reference to 'children as well as elderly people' suggests that the introduction probably lagged a few years behind the US. As shown by Bentley (2014), at its introduction in the late-nineteenth century, formula manufacturers were not exclusively targeting babies but also older people and 'invalids' because of the product's high nutritional content, in this case also mothers wanting to increase the production of breastmilk. It was sold in pharmacies, just as was the case in Sweden. Today, the clear, green glass bottles for Mellin's Infant Food are collectors' items and sold at high prices on the Internet.

The link between bottle-feeding and good motherhood was spurred by the idea of 'scientific motherhood' (Apple 1995, 2006), guided by medical advances and the detection of bacteria as the cause of major infectious diseases in the late-nineteenth century. Although it was generally considered that breast milk was the most hygienic baby food, infant mortality rates in cities were high and infant feeding became the keystone of the emerging specialty of pediatrics. Gradually, during the first decades of the twentieth century, bottle-feeding was increasingly established as natural, necessary and safe, thanks not least to the invention of the Indian rubber nipple (teat) that was boilable and thus possible to sterilize (Apple 1987).

The next big step in the development of processed baby food was driven by the discovery of vitamins in food in the early twentieth century. Until then advice manuals had not prescribed that children be given fruits and vegetables until two or three years of age (Bentley 2014). Canned goods had been available since the late 1800s, but were too expensive for most, and it was not until the 1920s that canned produce was manufactured in sufficient quantity to allow mass consumption and connections to be made between consumption and motherhood. The first commercial baby food on the US market was Clapp's Baby Food who produced a 'baby soup' already in 1921, a combination of beef broth, vegetables and cereals. But the real success of commercial baby food was the introduction in 1928 by The Gerber Company of its new baby food line of canned 'strained' (pureed) peas, prunes, carrots and spinach. As Bentley (2014) shows, conditions were favourable—advertising had become widespread, the cost of canned foods had fallen, and experts recommended the addition of fruits and vegetables to the infant diet—and Gerber's baby foods were quickly adopted by American consumers. The company soon added a line of cereals and within a few years introduced chopped produce and dinner combinations for older toddlers. Part of the success was the introduction of the iconic Gerber baby—tousled hair, bright eyes, rosy cheeks and round pursed lips—that featured in advertising campaigns and on labels: the epitome of the happy and healthy baby that has dominated the marketing of baby produce ever since. The illustration became so popular that Gerber adopted it as its official trademark in 1931 (Bentley 2014, pp. 20–26).

By the end of the 1930s commercial baby food had acquired mainstream status in the United States, thus solidifying the notion of the special category of food for babies that had not previously existed. By the post-World War II period, baby food had become a fully naturalized product in the US, thanks to widespread availability, persistent marketing campaigns, and a strategic alliance with pediatricians and childcare experts. In the 1960s and 1970s there was a veritable backlash towards commercial baby food as the inclusion of added salt, sugar, fillers and preservatives came under fire and studies showed that such additives might be harmful for infants. But baby food companies were quick to adapt and what Bentley refers to as the New Baby Food (2014, p. 143) emerged, focused on taste and texture, organic ingredients, the avoidance of allergies, and care and convenience for working mothers. Once again, commercial baby food became a normalized part of the baby's diet thanks to connections between medical science, health care, advertising and companies, this time also taking into account the views of the women's and consumer movements. New associations were made and commercial baby food was re-inscribed with the values of the time.

In Sweden, the introduction and normalization of processed baby food is closely linked to the development of the (still market-leading) company Semper. Semper has always been connected to science and food technology, including the new method of spray-drying milk invented by Swedish nutritionist Ninni Kronberg in 1933 and, in 1939, Svenska Mjölkprodukter AB (SMP) (Swedish Milk products, later Semper) was launched. In 1941 the concept of *välling* was first introduced: a mixture of skimmed milk powder, various flours, minerals and vitamins that is still today the staple for Swedish babies. In 1948 the company's own formula, BabySemp, was launched, advertised as 'the world's first industrially produced infant formula'.

The first jarred produce was not introduced until 1961, basically as a way to meet competition from another Swedish company, Findus (Fagerfjäll 2012). Findus (short for 'fruit industries') was a growth company with the mission to test new ideas from North America in the Swedish market (see Chap. 2). In the early 1940s, the company had a line of canned jams, beans and cucumber salads; in 1945 they pioneered frozen food; and in 1948 they launched the first canned baby food in Sweden. The premiere range consisted of artichoke, carrot and apricot puree. To meet competition, Semper introduced a range of jarred dinners for babies in 1961 based on ingredients such as lamb, ham, chicken, liver and ox tongue together with rice and vegetables, resembling the ideal Swedish family dinner of the time. Today Semper advertises a range of organic baby foods as well as taste and fun for babies and convenience and care for mothers, similar to developments in the US (see Fig. 4.1).

Taken together, the various accounts of the history of commercial baby food illustrate how the elements of processed baby food have been configured and normalized as part of parents' weaning practices at the same time as they make evident the continuous production, reproduction and collaboration between various actors needed for a practice to endure and exist over time.

Fig. 4.1 Assortment of Semper baby food from 1961. Source: Semper archive at the Centre for Business History, Stockholm

Commercial Baby Food Today

The historical account illustrates the way commercial baby food became a common part of weaning practices in Sweden and other western countries. The normalization process involves a continuous interplay between medical science, health care, advertising, product and packaging innovation, socio-economic changes and ideas such as 'scientific motherhood' (Apple 2006) and 'consuming motherhood' (Cook 1995; Bentley 2014).

Below, we develop our analysis by illustrating how commercial baby food products fit into contemporary weaning practices and thereby remain a normal part of weaning in contemporary Sweden. As we illustrate, the normalization of commercial baby food is conflated with weaning practice—a future-oriented project involving parents' emotions and visions of a healthy and weaned child sharing food, meals and mealtimes with the family (Brembeck and Fuentes 2017). Weaning practices involve a number of elements: parents, babies, food, cutlery, and meanings and ideals that are linked and organized in accordance with understandings and rules on weaning, including advice given by health authorities and cultural understandings of 'care' and 'good mothering'. But they are also held together by teleoaffective structures (Schatzki 2005), that is 'the ends towards which engagement in the practice are oriented' (Warde 2016, p. 40), namely babies eating solid food. Understanding how commercial baby food becomes meaningful and normal implies taking an interest in the organized arrangement of elements that are involved in weaning, as well as the performance of the practice and how they are linked to babies eating solids. Hence it involves exploring how commercial baby food fits in and re-organizes the arrangement of elements that keep weaning practice in place—what prescriptions they entail, how they are de-inscripted and how gaps or tensions are solved through re-inscription.

As Good as Homemade?

Across differences in age, educational level, employment and number of children, all parents we met during our fieldwork at the baby café used commercial baby food. Parents described how convenience food products were part of their daily care practices, used interchangeably with fresh ingredients and home-cooked food. The extent to which convenience products were used varied from parents mainly feeding their babies commercial baby food to parents cooking most of the meals and only using convenience products for travel, or as a backup for occasions where they had no time to cook. There was also obvious variation among meals, i.e. while one meal might consist of a canned ready meal, another could be a mix of ready-made and raw ingredients and yet another cooked from fresh ingredients. These findings parallel those of previous research describing how consumers make convenience products normal and appropriate parts of their everyday chores by integrating them in a mix of products. In previous accounts of consumers' everyday use of convenience food, meals stood out as heterogeneous rather than homogenous in terms of ingredients use. In parents' accounts, convenience food products and fresh food appeared to be part of the same continuum rather than absolute alternatives (see Marshall and Bell 2003; Moisio et al. 2004; Halkier 2016).

It was also obvious that all parents were aware of discourses positioning convenience food as inferior to homemade, but the general opinion was that commercial baby food was 'OK', at least now and then. Klara, a first time mother of an 8-month-old son, touched upon this tension: 'We do in fact cook most of the food. But then sometimes you are tired and cannot cope and then you buy something ready-made. Absolutely. We are all human'. Petra, a mother of three, argued for a relaxed attitude: 'My advice', she wrote, 'is not to fuss. If you do not want to make a lot of food yourself, it is OK to buy readymade. You should not feel shame for buying everything ready-made and not cooking yourself'. Klara's pardoning phrasing 'we are all human' and Petra's reference to 'shame' reinforces the idea that commercial baby food is morally ambiguous. Acknowledging the ambiguity of convenience products while at the same time justifying their use, our findings echo those of previous studies of convenience products and care practice (Carrigan and Szmigin 2006). Like previous studies of convenience food and mothering, our findings suggest an ambiguous understanding of convenience products. But as we illustrate below this ambiguity, although visible at times, could be 'handled' in practice (see also Chap. 7).

In the case of commercial baby food and weaning practice, a key aspect of this 'handling' was the apparent match between products' pre-scriptive forces and consumer de-inscriptions of the same. This was in turn linked to the products' abilities to help parents overcome the commercial/homemade dualism and to fulfil other ideals, needs and demands. Baby food companies actively inscribe their products with qualities of authenticity and quality. At times, the moralized discourse of convenience food is addressed head-on as in the case of Semper linking their product to ideals of 'real food':

> Some call it baby food. We call it real food. For babies. Naturally tasty food, cooked from the best ingredients. Nothing strange, no shortcuts and no additives. Real food—since we cook food using the best ingredients, with no preservatives and no strange additives. For babies—since our food contains just the right amounts of nutrients, size of pieces and good taste that suit a developing baby. (www.semperbarnmat.se/riktig-mat-barn, accessed 3 April 2015)

In addition, the baby food is given 'real' food names, e.g. 'Kalops' (beef stew), 'Moussaka', 'Coscous med kyckling och basilika' (couscous with chicken and basil). 'Real food' connotes ideals of a 'proper meal' that is linked to notions of domestic cooking 'from scratch', with skilful use of high-quality ingredients and care (cf. Bugge and Almås 2006). It is evident that the marketing text from Semper cited above and the many additional examples found on company websites are part of an effort to portray commercial baby food as equivalent to homemade food. As well as being 'real', the food products are also inscribed with other qualities central to weaning practice. Besides an awareness of discourses positioning convenience food as inferior to homemade food, parents were also highly aware of many other ideals pertaining to care, health and good parenting as well as specific material and temporal demands that often seemed to outweigh ideals of home cooking.

In the following sections we describe four ways of using or 'de-inscripting' commercial baby food that, despite the moral ambiguity of commercial baby food, came across as mundane and normal: 'getting it right', 'avoiding panic', 'feeding on the move' and 'getting the baby to eat' (see Fig. 4.2). These four ways of using commercial baby food were not mutually exclusive but occurred interchangeably over time. Moreover, they seemed to involve settings or situations where the prescriptive forces of canned and bagged baby food made a particularly good fit with parents' de-inscriptions of products, i.e. their reaction to and use of the same. These were situations where parents were particularly anxious to follow public health advice and '*get it right*', when the pre-scriptions of immediacy, flexibility and mobility had a good fit with parents' efforts to '*avoid panic*' and feed the baby '*on the move*'. Lastly, sometimes parents found it hard to make the baby interested in their home-cooked food and parents then described commercial baby food as a way to '*get the baby to eat*'.

Mundane use of commercial baby food	Materials	Integrated elements Meanings	Competence
Getting it right	*Välling* (gruel), porridge, Canned meals Babies + Feeders High chair Plates and cutlery	Proper food Nutritious and healthy Tailored for babies' needs Local and eco produced	Managing meal routines Adhering to public advice
Avoiding panic	Packaged meals and fruits Babies, Parents etc.	Easy to store Immediate food Stress-free	Feeding here and now Self-feeding
Feeding on the move	Cans and pouches Public feeding infrastructure Babies and parents	Mobility Less mess Manageable for adults and babies	Feeding here and now Self-feeding
Getting the baby to eat	Porridge, packaged meals Babies, Feeders High chair Plates and cutlery	Tasty, Healthy Relaxed Comfortable	Getting the baby to taste and eat food

Fig. 4.2 Ways of using commercial baby food

These four ways of using or de-inscribing baby food are all part of the practice of weaning where parents seem to react to and use commercial baby food in line with the product pre-scriptions. Although aware of the moral ambiguity of convenience products for infants, the parents interviewed described commercial baby food as a healthy, flexible and appropriate way to serve their baby food. It was described as a product used to handle the demands of flexibility and immediacy that characterized the weaning process while at the same time adhering to ideals of proper food and good mothering.

Getting It Right

All the parents we interviewed agreed that weaning was not a natural process that happened by itself. Instead it was a process of parents learning the skills of weaning and babies the skill of eating. In turn, this process required a specific arrangement of materials and competences, meanings and understandings that aligned with explicit rules and instructions as well as cultural ideals. The parents described weaning as a sequential process with several steps requiring specific foods and know-how. All made reference to the guidelines and instructions they received from baby health-care nurses. These guidelines seemed central to the way weaning was understood and performed. The guidelines prescribed a form of 'career-led' weaning (Keenan and Stapleton 2014) and took the form of a step-by-step programme. The food was supposed to be introduced gradually, starting with mixed food that can be swallowed without chewing at the age of 4–6 months, and later on meals with several ingredients and pieces to chew at the age of 6–8 months and complete dishes from around 8–12 months (Trädgårdh Tornhill 2012). The guidelines also recommend that parents should avoid salt before the age of 12 months when the baby's kidneys are more developed and avoid serving milk or yoghurt due to the risk of iron deficit.

In the parents' stories, weaning was a tricky business where the needs of babies in terms of nutrition, motoric challenges and intimacy had to be managed. Mirroring descriptions and language used in health-care advice, parents seemed to have a medicalized understanding of mothering and weaning (Apple 2006) that emphasized scientific knowledge and medical expertise while undervaluing the everyday experience and know-how of parents. 'Baby-food' and 'baby-feeding' are separated form 'ordinary' cooking and eating practices in a manner important to the use and understanding of commercial baby food.

The marketing material and packaging of commercial baby food clearly illustrates that product design prescribes a use that corresponds with and fulfils all the demands of the model communicated by the medical health authorities. The Swedish market for baby food offers a wide variety of products, including pureed fruits, vegetables and meals, categorized according to the age and developmental stage of the baby. The step-by-step weaning model is inscribed in the material composition of products (e.g. ingredients and texture) as well as the marketing and packaging (e.g. labelling based on age-categories).

The surface of packaging is also used to inscribe products with a set of values central to care. Expressions like 'high quality', 'carefully selected', 'nutritious', 'tasty', 'natural' are often used, making the food products appear as material anchors for health and safety and a vehicle for the distribution of medical as well as scientific competence. Offering a child commercial baby food was portrayed as a way to secure a healthy, nutritious and scientifically-tailored diet. Parents used commercial baby food accordingly, to make sure all the baby's needs were met. Using commercial baby food was a way to adhere to the 'rules of weaning' and make sure the baby received appropriate food in terms of ingredients (no salt or sugar), nutrition and texture.

Porridge stood out as a particularly evident case of commercial baby food use where the commercial product is understood as 'appropriate'. It seems that these products have a specific position in the weaning process and were used habitually even by parents who made an effort to make the rest of the baby's food themselves. Hence, convenience food products were categorized differently in parents' stories. While canned dinners were easily recognized as 'convenience food', commercial porridge powder that parents mix with heated water was not always acknowledged as belonging to the same category.

The emphasis parents placed on the guidelines and the step-by-step model prescribed by health-care advice makes marketing appear central to the normalization of commercial baby food. The on-going achievement of making commercial baby food normal seems to rely on the commercial products' pre-scription of the same model. Baby food companies design their products in a manner that materializes the very 'route' of career-led weaning prescribed in health-care guidance, making the use of the products seem appropriate.

Avoiding Panic

Caring for a baby is a time-consuming and physically demanding task, especially when the needs of the baby must be integrated with the time schedules and wants of other family members. The participants in our study describe how they devote time and effort to managing the feeding of a baby in relation to the mealtimes of the rest of the family. Babies need to eat often and sometimes irregularly (although most mothers made an effort to maintain a strict routine) and there were frequent descriptions of situations where food was needed almost instantly. Babies' frequent and immediate need for food fits well with the packages' prescribed servings of small individual meals. Commercial baby food was often kept as a backup for occasions when the baby needed to eat 'now' and there was no time for cooking (see also Chap. 5 on the temporalities of convenience food). Interviewed by Maria Fuentes, Ulla describes how she started to cook for her son but still uses commercial baby food:

Maria: So you still use jars as well?
Ulla: Yes, yes we do. Time is the thing really. You think you should have time to do things but he does not sleep that much. And he climbs up my legs so sometimes it is just not possible [to cook]. He runs the day really; well you know how it is.

The uses of convenience products as a way to manage time have been described repeatedly in previous research on the use of convenience foods (e.g. Warde 1999; Carrigan and Szmigin 2006). A similar way of using convenience food was described in reference to occasions when the baby was fed the same food as the rest of the family or home-cooked food was made especially for him/her but the baby refused to eat it for some reason. Commercial baby food was kept in store for those occasions to avoid not being able to make the baby eat. Commercial baby-food products facilitate haste, pre-scribing the instant serving of food. Although some companies advise against it, the food can even be served directly from the jar or pouch.

Commercial baby food products also pre-scribe ways of preparation that make 'cooking' with a baby more do-able (Halkier 2010). For example, *välling* and porridge packages invite one-hand preparation and serving while pouches suggest their use to be less messy and invite babies to eat with only minor assistance from an adult. Hence, commercial baby food packaging is designed to be easily manageable, for example, using only one hand while carrying the baby with the other or involving the baby in a different way in the feeding process.[2] Therefore, commercial baby food prescribes a way of cooking and feeding that gives parents room to manoeuvre the baby and his/her needs while feeding. Again, the analysis shows that convenience food offers a good material fit with the demands of weaning practice.

Feeding On the Move

Commercial baby food was also repeatedly described as a way to handle feeding when meals took place outside home. This is parallel to many other convenient ways of providing meals and eating 'on the move' (Halkier 2016). When asked to describe the content of their nursery bags, all parents reported canned meals and/or pouches of fruits and yoghurt to be a standard part of the equipment. Sigrid, a 30-year-old first-time mother of nine-month-old Saga, described how she used to 'keep a bunch' of pouches of food when she planned to be out all day, and was unsure of when and where meals would take place. Besides 'emergency food' she also described how Saga finds the products fun to play with after the meal. Anette, a 28-year-old first-time mother of a seven-month-old daughter, described a similar approach, always making sure to have pouches of fruit or yoghurt in her care bag to have something at hand if her baby gets hungry. She also underlined her efforts to bring food that is less messy and how she tries to spoon-feed her daughter when out while encouraging her to handle the food when eating at home. Ella describes how she usually keeps a baby-food jar or two in her nursery bag just to make sure she has something to offer her baby if the food she eats is not appropriate for her (for example, too salty or spicy). For Sigrid, Anette, Ella and other parents like them, commercial baby food becomes a way to solve the predicament of integrating the baby's needs for food with mobility. Babies eat often, they require special food, they generally need some form of diversion or amusement while eating and meals are often messy. Commercial baby food, offering instant food in colourful 'interesting' packaging, fits well with meals eaten outside the home when 'on the move'.

In contemporary Sweden, weaning is a dispersed practice in the sense that it takes place over time and in various places. Contemporary ideas on child development and child rearing encourage parents to feed the baby according to his or her own schedule (Brembeck and Fuentes 2017). In a recently published report, Semper describe Swedish parents as active and highly mobile and therefore in need of 'smart' products delivering convenience (Semper 2016). Weaning is not just something practised at regular mealtimes or at home and producers of commercial baby food design products that prescribe mobility and frequency.

The products are designed in small portion packages that enable mobility and frequent use. Jars and pouches are easily fitted into the mandatory nursing bag where parents keep diapers and napkins, extra pacifiers, drinking mugs and bowls and spoons for feeding. The products do not require cool storage or time-demanding cooking. Parents can easily have baby food at hand, at home or on the move, and the products are easily heated on the stove or in a microwave oven. Many cafés and restaurants offer microwave ovens for heating baby food that guests bring with them to feed their babies.

The material qualities of the jars and pouches fit with modern life, the material 'feeding infrastructure' present in homes, cafés and restaurants, and babies' needs. The dispersed character of the practice is one of the aspects that makes commercial baby food a good fit with the nexus of elements in weaning practice.

This is, however, not enough to normalize products. Mobility and frequency, or flexibility, if you will, are characteristics that can be ascribed to most convenience food of which many are deemed inappropriate (Halkier 2016). And like other convenience food products, commercial baby food is industrially produced for commercial purpose. There are many other forms of convenience food that are widely available and publicly used, for example fast food, but which are still not considered appropriate. To be appropriate and normalized in moral terms, commercial baby food products are re-inscripted in reference to the convenience food category.

Making the Baby Eat

A fourth way of using commercial baby food was described by parents who had problems motivating their babies to eat. Commercial baby food was described as a way to make the baby eat. Jessica, a 32-year-old mother of two children, describes how she tried to cook for her daughter:

> I tried but she did not want anything that I cooked myself. She did not even want potatoes. Ordinary potatoes. She almost vomits when I feed her something that I have cooked myself, so we buy food. It is the only thing that works.

Although wanting to make food herself, Jessica succumbs to the demands of her daughter and offers her commercially-prepared food. For Jessica and other parents who describe having problems motivating their baby to eat, commercial food was portrayed as a way to encourage babies and help them to acquire the skills of eating.[3] Commercial baby food was described as well-liked by the babies and most parents could account for one or several 'favourite' products that could be used to solve a tricky feeding situation. Few parents offered any reflections on why their babies liked or even preferred specific food products. Commercial products were on a few occasions described as tastier than homemade food. A closer look at the products offered on the Swedish market suggests that the dominating brands like Semper and Ella's Kitchen offer food with a natural rich or sweet flavour. Products marketed as 'tasties' for babies 4 months and older include, for example, pureed maize, pureed carrot or pureed fruits. Meals marketed as suitable for children 6 months and older often include spices like basil, garlic and other aromatic ingredients like tomatoes and cheese. The varied assortment offered also invites parents to sample various tastes and even try different tastes at the same meal.

The preference of babies was described as one of the primary criteria for their choice of baby food and parents seemed to frame the use of 'favourite' commercial food as a way of being responsive and baby-oriented. In these situations it was clear that the commercial-homemade ideal was subordinate to the ideal of being child-oriented.

A closer look at the commercial products and especially the packaging illustrates how they pre-scribe child-oriented feeding (Fuentes and Brembeck 2017). As previously described by Wagner (2013), this is particularly evident in the case of the food pouch that was introduced in Sweden by Ella's Kitchen. As Wagner (2013) points out, the soft, squeezable and often colourful food pouch shares many traits with children's toys. Parents described how the food pouch could be used as a distraction before, during and after eating. Compared to glass jars, the package is also safer for the child to handle and the spout pre-scribes sucking, a way of eating that is familiar to all babies whether breast- or bottle-fed (Wagner 2013). Babies are able to interact directly with the package, fulfilling the dual purpose of feeding and preoccupying the child.

Expected and Acceptable Convenience

The parents in our study described commercial baby food as an expected and acceptable part of feeding the baby. The parents' accounts demonstrate how commercial baby food as a form of convenience food has become incorporated within their everyday routines and food practices including the way shopping and cooking has been 'conveniencized'. The parents' stories occasionally disclosed an awareness of the moral ambiguity of convenience food as a category, but at the same time commercial baby food was described as a sound option used at home and in public.

The brief history of commercial baby food, together with the four ways of using commercial baby food, illustrates the 'normalizing work' required to make commercial baby food remain a normal part of weaning. The parents' descriptions show how the products made feeding more do-able (Halkier 2010) and offered guidance for an appropriate way of acting (Jelsma 2006). The products privileged a certain trajectory described by the parents who used the products to secure a healthy diet, solve emergencies, stay mobile and respond to babies' preferences. Commercial baby food was seen as a good way to conform to the weaning standards set by health-care officials and to manage and integrate weaning in the multitude of practices that make up everyday life. Hence, the marketing and design of products as well as the parents' actual use and understanding of the products were central to its normalization.

The link between normalization of convenience food and do-ability is visible in the German, British and Danish fieldwork too, and also corroborates findings in previous studies of convenience food (Carrigan and Szmigin 2006; Carrigan et al. 2006; Halkier 2016). This is particularly evident in descriptions of convenience products as a means to time-shift and solve emergencies (see also Chap. 5 on the temporalities of convenience food). British, German and Danish participants described convenience products of various sorts as a way to time shift and manage the sometimes unexpected flow of everyday life. For example, the British and German material includes accounts of storing convenience food in order to have something available to offer unanticipated visitors (with some participants referring to their food cupboards as nuclear bunkers or war-chests). In Denmark, parents also use convenience products to manage the sometimes-shifting flow of everyday life with children, such as ready-made buttermilk pudding, pita-breads and the Danish speciality, seen as convenient, rye-bread open sandwich: 'But you know, it's not that long ago, where we just had a fried egg on a piece of rye-bread' (Anja). Even the canteen case study illustrates the expected and accepted use of convenience food in case of unexpected events. One of the canteens featured a *Havarieregal* (a 'hazard' or 'emergency shelf') where canned ready-made foods were stocked in case something unexpected happened to the food they had prepared, making it unfit to serve to customers.

It is also important to note the discrepancy visible in the material from the various countries and cases. While Swedish parents describe commercial baby food as an expected and accepted part of any meal, the British and German descriptions address convenience food products in general and more often involve justifications for their use (see also Chap. 7 on the moralization of convenience food). This would suggest that commercial baby food, like the Danish meal-box schemes, represents a 'proper' way of making feeding convenient. The parents' stories suggest that commercial baby food products hold moral capacities (Jelsma 2003) to be used in the enactment of being a good parent and responsible consumer.

Commercial baby food offers a material arrangement of food provisioning, cooking and feeding that supports convenient *and* morally appropriate ways of acting. Products enable 'conveniencization' in terms of flexibility, mobility and less time spent on planning, shopping and cooking—but also in terms of promoting food conduct that parents understand as morally appropriate. Hence, in the case of commercial baby food 'conveniencization' involves making moral conduct more accessible (as also discussed by Fuentes and Fuentes 2017).

Commercial baby food is produced, designed and marketed to be 'convenient' but also 'natural' and 'nutritious' and sometimes 'local' and 'sustainable' (see also Fuentes and Brembeck 2017). Also, although commercial baby food makes weaning more convenient, its use still involves various forms of work. Parents have to learn about products, acknowledge issues such as sustainability and provenance and make choices that fit the age and preference of the baby. They have to heat and/or serve the food and feed it to a sometimes-reluctant baby. This work enacts the values inscribed through design and marketing and rescripts the food products, making them potential resources for the enactment of good mothering and responsible consumption (cf. Bentley 2014; Fuentes and Brembeck 2017). They offer parents a way to 'do the right thing', following the standard set by health-care officials and expressing good parenting via the consumption of 'good' food products. In this sense, commercial baby-food products were used as moral objects (Jelsma 2003) in the enactment of the 'good mother' (Bentley 2014) but also the 'active consumer' (Trentmann 2005).

The historical development of the Swedish baby-food market and the four 'normal' ways of using commercial baby food today illustrate the products' role as moral but also 'political devices' prompting the articulation of various interests (Hawkins 2012, p. 67) from baby health and child-orientatedness to sustainability. The market for commercial baby food products grew out of a continuous interplay between medical science, health care, business and advertising (Bentley 2014) and this interplay is still vital to maintaining its 'normal' status. The market for commercial baby food is, as illustrated above, enabled and shaped by various interests such as the recommendations from the National Food Agency, health care advice, and discourses on sustainability and provenance that are inscribed in products.

Acknowledging these various interests through the scripting work of product development and marketing as well as via the work of consumption, commercial baby food materializes what Hawkins describes as a form of 'political participation and citizenship' (Hawkins 2012, p. 80). Commercial baby food constitutes a mothering subject (see also Bentley 2014) that is able to 'do good' by combining the use of domestic technology, convenience products and other foods rather than cooking from scratch (Fuentes and Brembeck 2017). Although the products offer convenience, they also demand various forms of consumption work. With reference to the term 'micropolitics', Hawkins (2012) illustrates and discusses the way mundane products and specifically packaging that needs to be recycled, can disrupt the habits and self-cultivation of household members (ibid., p. 80). Similarly, commercial baby food enables and shapes the ways parents approach and understand weaning. Parents have to choose, combine and prepare food products but they do not have to imagine a suitable menu, pick out ingredients or cook. The 'conveniencization' of weaning implies a shift where skills and knowledge, regarding, for example, nutrition, meal composition and cooking, as well as the handling of parental concerns about food, are relocated from the consumption domain to commercial service-providers. This has an impact on the habits of households and changes the prerequisites for being a parent and a consumer.

Just as the twentieth-century introduction of food packaging made consumers develop new ways of evaluating the quality of products 'analytically' and 'from a distance' rather than trying or tasting a product (Cochoy 2005), so 'conveniencization' shifts the way people relate to food and food consumption. As more and more products involve various degrees of convenient-ness in terms of preparation and addressing concerns like ethics and health, the convenient food format becomes incorporated within their everyday routines and food practices. And it is not just complex convenient products but also the position of active yet 'convenient' consumers that become expected and accepted.

Conclusion

This chapter has focused on the normalization of commercial baby food to illustrate the on-going work of normalizing a morally ambiguous product category. Commercial baby food is commonly used in Sweden and although they are familiar with the moral ambiguity of ready-made food, many parents reported using commercial baby food without hesitation or guilt. Consumers successfully integrate these products in everyday care and feeding practices.

Narrating the history of commercial baby food as well as the contemporary use of the product, the chapter illustrates how the normalization of these products involves a continuous integration of various materials and methods. Through marketing and design, products privilege a certain trajectory that parents described as a way to secure a healthy diet, solve emergencies, stay mobile and respond to babies' preferences. Commercial baby food was seen as a good way to conform to the weaning standards set by health-care advisors and to manage and integrate weaning in the multitude of practices that make up everyday life. Hence, the marketing and design of commercial baby food as well as the parents' everyday use and understanding of the products were central to their normalization.

Against this background three observations on the normalization of commercial baby food can be suggested. First, the fact that Swedish health-care officials endorse the use of commercial baby food and describe it as equal to home cooking is of great significance. The positive attributes ascribed to commercial baby food in public material legitimize these products in a manner that separates them from other convenience food products. The quality of commercial baby food is also protected by legislation, banning the use of additives and taste enhancers thus protecting the products from many of the down sides of convenience food. Again, baby food is defined as something different from other forms of convenience food.

Secondly, the Swedish weaning standards and recommendations are inscribed in the product's design and marketing. This, along with the use of various claims of sustainability and provenance, gives products explicit and positive moral capacities as they support parents in their efforts to care for the needs and wants of their babies according to health-care standards and manage weaning in their everyday life. In this way, commercial baby food becomes a resource that can be used in the enactment of 'good mothering'.

Thirdly, and following the two previous observations, commercial baby food products carry political capacities, enabling and shaping the way parents approach and understand weaning. Parents do not have to make up meals, shop for individual ingredients or cook, but they still have to combine and prepare food products. This conveniencization of infant feeding implies a shift where skills and knowledge regarding, for example, nutrition, meal composition and cooking, as well as the handling of parental concerns, are relocated from the consumption domain to commercial service-providers. At the same time, other consumption tasks are added. These shifts have an impact on the habits of households and change the prerequisites for being a parent and a consumer. Commercial baby food, then, works as an arena for an active and reflexive consumer-mother (parent) who carefully navigates the offerings of the marketplace. Together, these observations illustrate the centrality of normalization in the century-long conveniencization of commercial baby food in Sweden, in terms of both becoming and remaining an expected and accepted everyday practice.

Notes

1. The Swedish text reads: 'För att barnet ska få i sig alla de viktiga näringsämnen som finns i fisk, inte bara DHA, bör föräldrarna i första hand uppmuntras att lära sitt barn att tycka om fisk. Ofta finns det någon fiskrätt som kan uppskattas. En del föräldrar tycker att det är svårt att tillaga fisk och kan behöva tips om hur snabbt och enkelt det faktiskt kan vara. Det finns också många bra halvfabrikat och färdiga fiskrätter att köpa' (Livsmedelsverket 2015, p. 27).

2. Examples of cooking with one hand while a mother balances a small child on her hip were also found in the UK (cf. Meah and Jackson 2017).
3. The British material also includes examples of parents who resorted to convenience food when their children refused to eat other kinds of food, using the argument that 'at least he's eating something'.

5

The Temporalities of Convenience Food

Introduction

This chapter explores the multiple temporalities associated with convenience food. As Shove et al. argue in their collection of essays on *Time, Consumption and Everyday Life* (2009, p. 2), time is about coordination and rhythm, but also involves material, emotional, moral and political dimensions—all of which are relevant to our understanding of the temporalities associated with convenience food. We begin by discussing changes over the *longue durée* which help explain the rise of convenience food, including the growth of female participation in the labour force, the socio-technical changes associated with the development of the 'cold-chain' in food manufacture and the growing use of domestic refrigeration and microwave cooking. We also discuss generational changes in attitudes to food, such as the post-war ideology of thrift and the suggestion that younger generations are more profligate in their attitude to food and less skilful cooks than their parents or grandparents (an assertion that we challenge in Chap. 8 on convenience and skill and in Chap. 9 on the health and sustainability of convenience food). We discuss changes over the life-course that help explain why different types of convenience food

P. Jackson et al., *Reframing Convenience Food*,
https://doi.org/10.1007/978-3-319-78151-8_5

are more popular at some phases in life than at others, particularly among young, single men and then again among older people, living alone.

Above all, though, this chapter focuses on the way convenience food helps solve what Alan Warde called the intransigent problems of scheduling in an increasingly de-routinized society (Warde 1999). We discuss Warde's argument about the benefits of convenience food in time-saving and time-shifting, as reflected in marketing claims about the way convenience food offers 'meal solutions' for time-pressed consumers, fitting in with the routines and rhythms of people's daily lives. Drawing particularly on the ready meals case study, the chapter extends these ideas concerning the relationship between (various kinds of) convenience food and different temporalities including tactics and strategies (such as 'stocking up' in anticipation of future needs); short-term practices (such as treats and rewards); longer-term horizons (such as ideas that associate convenience food with being 'modern'); and the implication of different kinds of food in memories of other times and places. We also look in more detail at the temporalities associated with the practices of shopping, cooking, and eating. Following up on our empirical evidence regarding the consumption of convenience food, we end by offering a systematic conceptualization of the temporalities involved in terms of transformations, rhythms and timings.

The *Longue Durée* of Convenience Food

In a celebrated article on history and the social sciences, Fernand Braudel famously drew a distinction between three historical temporalities: the time of events (told in 'the headlong, dramatic, breathless rush of narrative'); the time of conjuncture (involving slower-moving economic and social cycles); and the time of structural change, which he referred to in terms of the *longue durée* (Braudel 1958, as summarized in Ross 1995, pp. 188–189). Applying these ideas to the history of convenience food, we might think about the centuries-long impulse to preserve food, initially to meet the needs of armies on the move and then for wider commercial use. We might reflect on the industrial processes through which food became increasingly conveniencized, initially through canning and bottling and, later, through the development of 'cold-chain' technologies and other ways of prolonging the shelf-life of food, for which terms such as 'industrial freshness' have been coined (Freidberg 2009). We might consider specific innovations such as the development of the frozen American-style TV dinner, introduced by Swanson's in the 1950s, or chilled ready meals, developed by British companies such as Marks & Spencer in the 1970s. There are, then, numerous temporalities at play in the 'conveniencization' of food before we begin to consider the time of events and the use of convenience food in helping to juggle household schedules and the coordination of domestic life.

Kristin Ross (1995) describes the introduction of various domestic technologies in France in the 1950s, some of which (such as refrigerators) were directly involved in the conveniencization of food. Other innovations (such as washing machines and vacuum cleaners) provided benefits in other areas of household management. According to Ross, home economics textbooks and women's magazines enthused about the introduction of Taylorist principles to the organization of domestic life:

> Whole articles were devoted to the arrangement of appliances in the ideal kitchen or laundry room: the housewife should be able to proceed from one to another in assembly-line fashion without retracing her footsteps … No longer was a common-sense response, or the vague memory of how one's grandmother performed a task, sufficient—experts must be consulted, precise timetables kept to. By reducing the difference between the home environment and the workplace, house-work began to look more like real work—but not to worry, for the domestic appliances, essentially middle-class replacements for the nineteenth-century servants, the new *amis de femme*, were there to lighten the load. (Ross 1995, p. 102)

There has, of course, been much discussion about whether such labour-saving technologies did, in fact, reduce the amount of time spent on domestic work, given the need to clean and maintain such devices and their effect in raising standards of domestic cleanliness.[1] As this realization dawned on fictional characters such as Celine in one of the sources that Kristin Ross dissects (Christine Rochefort's *Les stances à Sophie*, 1963), Celine reflects that, for all the modern conveniences with which she has surrounded herself, she will still have to worry about what her family will eat every day for the rest of her life. Her sense of desperation is palpable as she concludes:

> And look at the amount of time wasted on that insane activity, one that has to be done again the next day, and the day after, and every day, and to think that there are 365 of them in one year alone and that we don't know how many years there are, and that on every one of those days the question will be asked and must receive an answer. (quoted in Ross 1995, p. 104)

This was the social context into which convenience foods were introduced and the domestic dilemmas that they were designed to resolve.

Before we leave the longer time horizons described by Braudel, it is worth considering the changing temporal patterns of domestic life into which convenience foods were designed to fit. This is nicely captured in work by Dale Southerton and others (Cheng et al. 2007) on changing meal-times in the UK (see Fig. 5.1). Little more than a generation ago, there were three distinct peaks during the day at which most British people ate their meals: breakfast around 8am, lunch around 12 noon–1pm and dinner around 5–6pm, with a further less pronounced peak around 10pm when some households had supper.[2] By 2001, this pattern

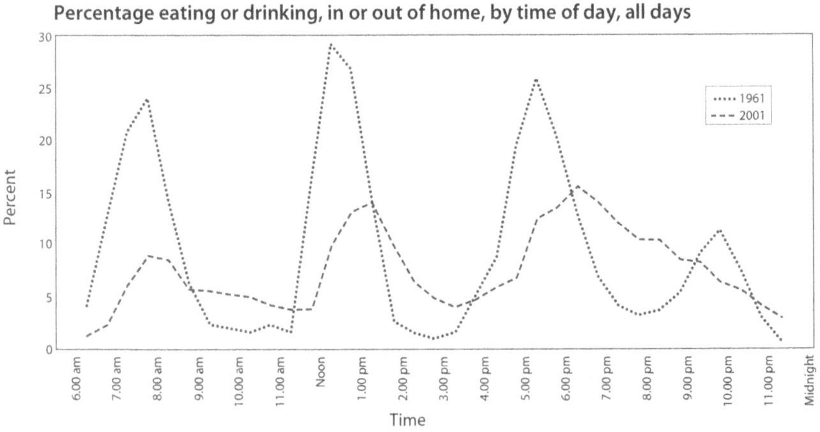

Fig. 5.1 Changing UK meal times, 1961–2001. Source: Cabinet Office (2008)

had changed dramatically with a much 'flatter' temporal distribution of meals taken throughout the day, including more frequent smaller meals, sometimes referred to as snacking or grazing (Warde and Yates 2017). These changing patterns have generated a heated debate about the alleged decline of the family meal, when family members all sit down together to eat (Murcott 1997; Jackson et al. 2009). Here, however, we focus on the role of convenience food in enabling or constraining the observed changes.

Practice theorists describe these processes in terms of the way people are enrolled into new practices, how they become entrained to them through processes of normalization and habituation, and how established practices may be relinquished as new routines become established (Shove et al. 2012). The language of enrolment and entrainment reinforces the idea of how shifting practices (e.g. from walking or cycling to driving) involve a series of institutional arrangements and socio-technical infrastructure that cannot be reduced to individual 'consumer choice'. Such is also the case with domestic cooking and eating, which Warde (2016) describes as a 'compound practice' because of the way it involves a range of subsidiary practices (such as shopping, storing, heating and disposing of food) as well as its complex inter-connection with other practices (such as work routines and leisure-time pursuits), all of which have to be managed in order to produce efficient cooking and eating practices.

Practice theorists also make a distinction between habit and routine, the former associated with predictability and regularity, the latter with a familiar sequence of activities that recur with a relatively predictable periodicity. But, as Southerton (2012) argues, the distinction is often obscured in daily life 'as many everyday actions are performed with a high degree of recurrence, periodicity and a degree of predictability; much action appears to be performed without reflexive deliberation; and many such actions are culturally shared, whether across whole societies or social groups' (ibid., p. 340). Southerton also warns that what might appear to be habitual actions through the patterning of practices might mask different dispositions towards the performance of those practices (ibid., p. 346). These arguments certainly apply to many aspects of cooking and eating as we explore below.

Domestic routines, such as cooking and doing the laundry, pose particular challenges in terms of scheduling and Shove (2003) suggests that the use of domestic technologies such as freezers and washing machines may, paradoxically, *increase* rather than ameliorate these problems, requiring greater coordination among household members. Shove makes the following argument: 'Initially referring to ease of use, then to saving time, convenience is now associated with the capacity to shift, juggle or reorder episodes and events' (2003, p. 170). Shove insists that 'convenience' should not be attributed to specific technologies such as freezers or microwaves as convenience is derived from how those technologies are articulated together and used in practice (ibid., p. 178). She also suggests that modern lives have seen an erosion of shared temporal ordering, within and between households, and a loss of collective coordination within the working day (ibid., p. 174). In busy modern households, the protection of 'quality time' is often held at a premium which the use of convenience food can help deliver. How this is achieved is discussed in more detail in Alan Warde's (1999) pioneering work on time-saving and time-shifting.

Time-Saving and Time-Shifting

Warde's argument about the space and timing of convenience food is underpinned by his argument that 'convenience' is not just a set of properties attached to specific types of food but is also a matter of social context (Warde 1999, p. 519). He argues that convenience food is a response to the de-routinization of everyday life and is about *timing* (scheduling and coordination) rather than *time* itself (ibid., p. 521). In contrast to media reports that emphasize the drudgery of cooking and the way convenience food enables a *saving* of time, Warde argues that convenience food should be understood primarily in terms of the *conceptualization and manipulation* of time. Warde quotes an article from *The Guardian* newspaper (6 December 1968) which reported that 'No one would deny the drudgery, the time-wasting, the monotony, that has been removed by convenience foods' (1999, p. 520). Pointing to the use of devices such as freezers and microwaves, he contrasts innovations which reduce the amount of time needed to accomplish a particular task with those that permit alterations to their timing in terms of scheduling or sequencing (ibid., p. 522). So, for example, freezing left-over food enables it to be consumed at a later time, though it may take longer to defrost and reheat than to cook it prior to freezing.

Warde suggests that convenience food can assist time-pressed consumers reduce the anxiety associated with organizing daily activities such as shopping and cooking, allowing them to store food that can be prepared quickly as and when it is needed. Similarly, he argues that women's increased participation in the labour force did not simply reduce the amount of time available for domestic tasks such as shopping and cooking, but that it imposed additional problems in terms of scheduling and coordination with other household members 'as the imperative to organise time, once the domain of industry, has invaded everyday life, eroding routines and compelling a significant proportion of people … to embark on scheduling strategies' (ibid., p. 524).

Warde's analysis is supported by arguments about the 'time squeeze' faced by modern consumers, discussed by Southerton and Tomlinson (2005) who argue that the accelerated pace of daily life is not just about a shortage of time (measured in terms of its volume) but also about coordination (scheduling) and time-allocation (among different tasks). Southerton and Tomlinson argue that UK time-use data reveal that people are not working longer hours, in aggregate, but that around three-quarters of those surveyed felt somewhat, fairly or very pressed for time. Professional and managerial groups reported feeling more time-pressed than other occupation groups and women were much more likely to feel time-pressed than men. These pressures are likely to have grown with the increasing flexibilization and informalization of work routines, as de-routinization makes coordination harder. Besides the 'dual burden' of paid work and unpaid domestic labour that many women carry, Southerton and Tomlinson's research raises questions about the distribution of obligations and responsibilities, particularly at key points such as the beginning and end of the school day when the juggling of multiple tasks can be particularly demanding for parents of school-age children. Convenience foods such as ready meals can be seen as a response to these temporal pressures, though its use is highest among single men and older people rather than among working mothers for whom it poses particular issues in terms of the normative values that are associated with 'feeding the family' (see Chap. 8).

In the following section, we use insights from the literature reviewed above to tease out in more detail the multiple temporalities of convenience food. We discuss both long-term and short-term changes, and the personal and collective temporal rhythms associated with the use of convenience food. We do so with reference to four temporal registers: routines, generational change, life-course changes, and modernity and memory, highlighting the intricacies of the temporalities involved. Secondly, we offer an analysis of these temporalities in terms of three elements of the compound practice of eating (shopping, cooking and eating). Finally, we make a more conceptual argument about the different temporalities observed, involving the transformations, rhythms and timings of the practices associated with convenience food.

The Routines and Rhythms of Everyday Life

Having reviewed previous research on the temporalities associated with convenience food, we turn now to our own data, drawing particularly on the case study of ready meal consumption in the UK and Germany. Here, we find many examples of the way that convenience food is used (often in combination with other kinds of food and modes of cooking) to fit in with our participants' multiple, sometimes conflicting, time commitments. Speaking specifically of ready meals, James (a 49-year-old professional), acknowledges that although these are not something he would choose to eat if it could be avoided, he felt that they have a place 'when time's against you'. As a single father of two teenage children who was trying to set up his own business, James felt particularly 'up against it', but this was no less of an issue among couples, even those without dependent children. For example, Edward and Deborah (a couple in their 40s, who are both employed full-time), spoke about how they try to make their shopping, cooking and eating 'fit in round the week', including their negotiation with household chores, gardening and pet-care. Deborah explained that she would like to make scones, rather than buying them, but no longer had the time: 'I love cooking. I love baking. I love the whole lot but I just don't have the time'. Even among participants who were no longer engaged in paid work, cooking and eating were fitted in around a range of other activities and obligations, including physical exercise and care of grandchildren.

Another UK participant, Tameka, a 28-year-old single mother with a young child, works part-time and lives on a social housing estate remote from any large supermarkets. Having no transport of her own, she shops online and opts for delivery at what might be considered an 'inconvenient' time—late in the evening. She does this because it is cheaper than peak-time delivery. She likes to cook but admits that 'I don't always have the time', particularly on a Wednesday, when she works a full day and then has to collect her daughter from the childminder. Although she reports that she tries to cook the night before, she does not always manage this. One Wednesday evening she video-recorded herself preparing a meal using a range of frozen 'convenience' items, including baby carrots, potato waffles, peas and veggie fingers. She explains to the camera:

> So it's Wednesday, normally I try to cook on Tuesday night so that I don't have to cook on Wednesday due to the fact that I do 8 hours at work. When I come home I don't want to cook. Anyway, today I'm doing something real quick because I don't have the patience to stand and cook as I'm so tired … Today I've got [lists items], very convenient and very quick to make before bed … So quick and easy as I get home just after 6 and really and truly my daughter needs to be in bed quite soon, so, if I'd cooked I'd just be warming food but, as I said, I haven't cooked, so at this point, this is the quickest thing I can do.

While Tameka enjoys cooking when she has the time and the resources, she uses convenience foods in a variety of ways. Sometimes it is to deliver a meal quickly before getting her child to bed; on other occasions, she will use items such as tinned kidney beans to create a 'traditional' Caribbean dinner of rice and peas as it eliminates the need to remember to soak dried beans the night before. Tameka employs a mixture of batch cooking 'for at least two days' and convenience food 'to cut out time [laughs] or eliminate long processes'.

In Germany, too, participants reflected on the way they combined 'convenience' and 'fresh' foods according to the time available throughout the week. Here, for example, is Nicola, a married mother of two children, in her mid-40s, who works full-time, describing her weekly cooking routine:

> We do use ready products, as you have seen, in our house, especially when it has to be quick during the week. But, at the weekend, I generally cook 'fresh' [from scratch]. And it tastes better, we think … also the children don't really know any different … Like at my parents-in-law, there's sometimes mashed potatoes from a bag [pulls a face to indicate disgust and laughs], er nobody likes that … Well, I do try to cook fresh during the week too, but I don't always succeed. Factor: time.

In this extract we see a typical example of 'good intentions' (to cook from scratch), thwarted on some occasions by a lack of time. There is also a telling combination of practical reasons for the choices made and a moralized language of distaste for certain kinds of convenience food, associated here with her mother-in-law's use of products (such as instant mashed potatoes) which, we are told, 'nobody likes'.

The routines of domestic cooking were also caught up in the rhythms of other social practices, dictated by available transport infrastructure and household technologies. As such, the location at which 'conveniencization' was delivered varied according to domestic circumstances, occurring at the planning stage in some instances (among meal-box users for example), at the point of purchase (where chilled or frozen foods might be purchased in anticipation of future use), at the preparation stage (where ready-prepared vegetables might be used to save the time and labour involving in peeling and chopping), at the cooking and eating stage (where micro-waved food might save cooking time or enable different family members to be fed at different hours) or at the cleaning-up and disposal stage (where prepared foods might be preferred as they are less likely to lead to waste than large quantities of raw ingredients that might go off before they can be consumed). Storing, freezing and reheating leftover food was also a common strategy, deployed for a variety of reasons including the need to cater for family members with different taste preferences or to accommodate a variety of incompatible eating schedules. So, too, did food-related practices such as procuring and preparing food have to be fitted in around other (non-food) practices—shopping on the way home from work or visiting a farmers' market while on a weekend leisure trip.

Other food-related routines occurred over a shorter timescale such as the use of treats and rewards by many of our participants. Here, the principle involves the 'right' to eat unhealthy foods such as chocolate or salty snacks, or to have a glass or two of wine, because it has been 'earned' by previous hard-work or well-disciplined dietary 'good behaviour'. This might involve a period of self-indulgence following a period of dieting or a regular Friday-night treat such as 'a cheeky take-away', where the antinomies of health and indulgence, economy and extravagance, described by Warde (1997) are temporarily reversed. So, for example, on an accompanied shopping trip with one of our German participants, Susanne allowed her daughter a plastic box of Baby-bel cheese as a treat but did not allow her a box of donuts. Susanne's family also always have frozen pizza in the fridge as 'a quick option', to which they added a few extra toppings and maybe a salad.

Treats and rewards are caught up in the discourse of 'healthy eating', which some participants deliberately repudiate as in the case of Tony, a middle-aged professional man who has recently separated from his wife. He declares: 'Pot Noodles are my secret vice … and I did have a couple of meatballs for breakfast this morning [laughs] as a little treat, it sounds terrible'. Tony describes his use of ready meals from Tesco's Finest range as 'like a mini treat' and uses humour to counter the discourse of shame and blame that is commonly associated with dietary advice (see also Chaps. 8 and 9).

Convenience foods can also be used when household routines are disrupted by the arrival of unanticipated guests. Many of our participants liked to have a store of convenience foods set aside for such eventualities. Even those like Ted, a 71-year-old retired professional, who says that 'convenience isn't a term I think of very much', kept a store of 'convenient' ingredients, including frozen peas and leaf spinach, and tinned tomatoes and pulses, while his wife, Laura, bought jars of artichoke hearts and sun-dried tomatoes to add to her lunchtime salad.

Many of our participants mixed convenience food and cooking from scratch. For example, Gloria reported how cooking and eating practices in her household often combined raw ingredients with pre-prepared or processed elements, including jars of cooking sauces which she might embellish by adding extra garlic. Gloria's husband, Jack, was particularly interested in experimenting with cooking and in video footage recorded by the couple we see them trying out a lasagne recipe in which courgette strips are substituted for lasagne sheets. Consulting the recipe card as she goes along, Gloria substitutes various fresh ingredients for those which she prefers or has in stock. The recipe indicates that a combination of cottage cheese and eggs should be used to make the sauce that will bind the ingredients together. In the absence of the former, Gloria departs from the recipe by using a jar of lasagne sauce. Aping a popular children's television show, she announces: 'Ta da! Here's a jar I prepared earlier'.

The cooking sauce is one of many items found in the household's larder and freezer which provided reassurance in being prepared for any eventuality. Indeed, revealing a cupboard full of convenience foods including breakfast cereals, Christmas pudding, jars of curry sauce, cartons of custard, tinned tomatoes, cup-a-soups, sandwich spreads, tinned spaghetti hoops, pasta and preserves, Jack declares: 'That's just in case we have a nuclear disaster, we'll be alright' (see Fig. 5.2).

Generational Change

As we shall see in our discussion of sustainability and waste in Chap. 9, there are significant generational differences in attitudes to food, with an older generation who grew up with memories of war-time food rationing (which ended in the UK in 1954) espousing the virtues of thrift, quoting maxims such as 'make do and mend' and 'waste not, want not'. They often disparage the apparent wastefulness of younger generations whom they consider less careful in their attitude to food and more inclined to waste it. This generational difference comes through in attitudes to use-by, best-before and display-until date labels, with younger generations despairing over the older generations' apparent lack of concern for the safety implications of eating out-dated food, while the older generations criticise younger generations for slavishly following labels, leading to needless waste, rather than trusting their own judgment.

Fig. 5.2 Gloria and Jack's emergency food supply. Source: © Angela Meah

Food campaigners also lament the apparent demise of cooking skills among younger people (an assumption we critique in Chap. 7), based on the view that members of older generations learned cooking skills from their parents (usually mothers)—a link which they feel has now been broken. This view is often based on another assumption, that parents and grand-parents were better cooks than their children. This idea was frequently challenged by our participants who recalled the bland food, limited culinary repertoire and over-cooked vegetables they encountered in their parental home (cf. Meah and Watson 2013).

In earlier work on consumer anxieties about food, Meah and Jackson (2016) compare experiences from several generations of the same families. For example, Sarah (aged 87), recalls her life in a rural area in northern England where her large family ate a lot of tinned food: 'Tinned peas and stuff … so I'd open a tin of beans and mince or whatever, erm, and make a meal out of that'. She remembers buying tins of Spam from the travelling shop as well as eating lots of potatoes and cabbages from the garden. She talked about feeding a large family on a tight budget and how 'you've got to create short-cuts all the time'. This is not what the discourse of generational decline leads one to expect in terms of the shift from scratch cooking, using fresh ingredients, to an over-reliance on convenience food, or the idea that culinary short-cuts are a relatively recent concern. Sarah's daughter, Jenny (aged 54), confirmed her mother's recollection, acknowledging that 'We used to have a lot of corned beef' (ibid., pp. 519–520).[3]

Life-Course Changes

Our data include many examples of how the use of convenience food varies with changes in the life-course, particularly at key moments such as the birth of children or retirement from paid work, as time becomes more or less available. While Meah and Jackson (2017) provide a detailed account of how parents' practices can shift even within a relatively short space of time during infancy, in the present study, grandparents also indicated how their own provisioning practices had to be adapted to the needs and requirements of visiting children.[4] In the UK, for example, Laura and Ted list the items purchased 'exclusively' for their young grandchildren. Laura mentions pizzas, chorizo and fish-fingers, to which Ted adds 'those little yogurts' and Laura continues: 'And meatballs I have bought for the kids ... beans, tomato, tins of tomato soup'.

Life-course changes do not, however, revolve exclusively around the needs or demands of growing children. Tony, for example, reflected on how the gradual demise of his marriage was manifested in the family's provisioning practices. Of these, he reported:

> The whole way the relationship was going, we were having lots of take-aways, not cooking properly [pause] became a bit of an odd lifestyle, the family life became a bit odd, so making stuff for [his daughter], but then just getting curries, takeaway curries and that sort of thing.

Following the break-down of the marriage, Tony moved into his own place when—contrary to the stereotype of the helpless single man having to fend for himself—he established greater routine in his provisioning practices, relying less on eating out and take-away food. Tony explains that this was partly 'because I like coming home more [laughs] in some ways', but was also financially necessary because he had to be more careful with his money. He explains that 'if I'm home I will cook something', simultaneously pointing out that 'since I moved out [three months earlier] I've only had about two takeaways, which is not much really'.

Our German data include several examples of participants who found they had more time to cook following retirement. In some cases, such as Petra (a retired development-aid worker), this resulted in her cooking dishes for home freezing, ready to defrost and reheat when she had guests to entertain, including meals that reflected her experience of living and working abroad.

Modernity and Memory

Returning to the longer time horizons with which this chapter began, we reflect on two other temporalities with which convenience food is associated. The first is the relatively intangible notion of how convenience food is caught up in the idea of 'being modern' and the conventional distinction that is often drawn between modernity and tradition. Many convenience foods were introduced as a modern, time- or labour-saving, alternative to 'traditional' ways of cooking and eating. While it is now common to disparage 'convenience' as involving an abrogation of 'care'—see Warde (1997) for the original formulation of this culinary antinomy and Meah and Jackson (2017) for a counter argument—the marketing of convenience food was often much more bullish about the alleged benefits of 'modernity' (as in the earlier quotation from *The Guardian* about eliminating the drudgery, wasted time and monotony of cooking through the use of convenience foods). The use of commercial baby food to replace breast-feeding was often represented as a modern practice, endorsed by the principles of 'scientific motherhood' (Apple 1995), and the marketing of many convenience foods has often poked fun at traditional alternatives (as was the case with the advert for instant mashed potatoes discussed in Chap. 2).

The fondness with which many of our participants recalled the advertising of their favourite convenience foods (such as Bisto gravy granules or Heinz tomato soup) also suggests that these apparently modern foods are able to evoke strong emotions and powerful memories in ways that might usually be thought to attach to 'heritage' foods, associated with artisanal production methods, following 'authentic' recipes (cf. Sutton 2001; Holtzman 2006). So, for example, despite his antipathy to most kinds of convenience food, retired British academic Ted kept tins of cassoulet and confit de canard in his store cupboard as reminders of happy holidays spent in the south of France.

There is also a sense of nostalgia about the way some of our participants described their meal times and the importance of eating together even if the ingredients would normally be described as convenience food. So, for example, Ken (an elderly rural participant who grows his own fruit and vegetables) and his wife Val talked about their insistence on eating at the table, acknowledging that 'a lot of families don't do that'. Apologising for the use of convenience food ('fruit out of a tin, I'm afraid'), they insisted on the virtues of eating together as a family:

> Your family comes together, not so much the husband and wife, but the kids. Er, I think if you've got the kids round a table, erm … I don't know, it's just an occasion when you're with them and they're not on their iPads or iPhones or the television or whatever … You've got their full concentration.

The contrast between such family 'occasions' and the use of modern devices such as iPads and iPhones reinforces the contradictory role of convenience food in promoting moments of family intimacy and mutual care. Likewise with products such as 'Very Easy Garlic', they maintained, 'it's so easy and it stores well and it's there and it's ready to use', providing a 'modern' (convenient) alternative with less smell and mess than fresh garlic. By contrast, Ken and Val felt that frozen chips were an inadequate substitute for 'real chips' ('I'll cook some real chips for you some time.... It's a bit of a performance but it's worth it').[5]

Convenience, Time and the Practices of Shopping, Cooking and Eating

Having discussed changes over longer and shorter time-scales and related dispositions towards convenience food, we now turn to a more detailed account of the temporalities involved in the practices of shopping, cooking and eating. Southerton (2012) distinguishes three temporal dimensions in relation to consumption practices: time as a resource (a finite amount of time per day, week or year), the temporal demands of practices, and the configuration of practices through personal and collective rhythms. The latter two elements are in a recursive relationship whereby the performance of a certain practice will take up a certain amount of time whereas the same practice is made to fit into the overall rhythms of daily life. In what follows, we discuss how this recursive relationship plays out in relation to convenience food, with a discussion of shopping, cooking and eating.

Shopping

There are great variations in terms of when, how often and where people do their shopping. The practice of shopping involves the time needed to get to the shop, to do the shopping and to bring the goods home. It also involves planning and decisions about future uses of the food items purchased. Making shopping more convenient may mean spending less time shopping or at least reducing the time-consuming elements of shopping. For example, one of our respondents, Melanie, describes how she and her partner Tom decided to switch to online shopping and home-delivery for convenience:

> Once a week he would do the shopping on his way home from work … So he would finish work at 5pm, go to Asda, it'd be really busy, he would shop for 45 minutes … and then he'd have to drive the rest of the way home and he would arrive home with all the shopping … And he'd be really grumpy, the evening would be kind of, eaten up … It was always this like … 'urgh, we gotta do the shopping today' kind of thing … I'm more of a planning sort of person, so the online stuff suits me, because I can really plan what I want and then get exactly what I need … And we'd have to take the car out especially on a Saturday or a Sunday and go to the supermarket and the supermarket would be really busy on a Saturday or wouldn't have stuff in on a Sunday etc … so we ended up doing it online. It costs about £1 or £2 to have a delivery, and we decided that's definitely worth it. For the convenience of not having to spend the time at the supermarket.

For other participants, convenient shopping is less about the time spent at the shops than about coordinating the occasions for shopping and the temporal rhythms of the household. One of our German participants, Thomas, made an explicit argument for the convenience of going shopping more frequently:

> We do it spontaneously. There are a few patterns that occur often. Susanne likes to keep the weekend free of the shopping stress. And she doesn't work on Fridays … usually … But we're not the kind of people who do that big weekend shop, planned, well organised etc. so you know what you need to buy … generally it's for the weekend. Plus maybe some stuff … but then usually during the week, then you need to go again.

Thomas goes on to talk about how longer-term planning may result in food being wasted:

> When we try, we attempt to, say, plan all through the week or at least until Thursday, what we'll eat, eh, eh, that results in us throwing away a lot.

> Because it never works out as planned, we're too undisciplined for that, because then, like, we go out for another meal, or we're fed up … because when you plan ahead, you plan more complicated things than you're then actually gonna cook … or there's more leftover … if you want to buy something at least a little bit fresher then, then, … we already throw away much too much anyway … and, so, that's why it doesn't work really with the planning ahead. So usually we still have something for Monday and Tuesday and then, eh, it's either Susanne in the evening or, me on the way home, to drop by somewhere or I set off again and drive to the shop later.

Apart from the continuum of planned shopping practices designed to take up as little time as possible, on one extreme, and frequent unplanned shopping trips to allow for spontaneity, on the other, convenience also comes in when temporalities besides shopping are considered. In our data, we find evidence that some respondents try to buy things they know will be easy and convenient to prepare or that will be easy to store and last as long as possible. As Carl, who lives in a student flat-share, says:

> I always make, kind of quick things … I look for offers and then I see what I could cook with that.

In this case, convenience foods are valued for their speed and cost, and for the spontaneity they afford (when shopping and cooking).

Cooking

Following Warde's lead, using convenience food and making food preparation convenient is not just about time-saving but also about time-shifting to accommodate people's increasingly de-routinized daily schedules. Fridge, freezer, microwave and electric stove are obviously of great importance in the practice of time-shifting (Shove and Southerton 2000). But it is not just convenience food which circulates through chilling and heating devices but also self-made meals which are prepared in greater quantities, ready-to-use conveniently at a later time. Respondents from our German fieldwork explained that they are saving food for later occasions, both planned and unplanned:

> So this is home-made hummus … and that's Rendang I made (beef in a coconut sauce). It's practically ready. This is very good when we have guests, then … I have several tubs, I can see how much I need. (Petra)

However, the temporal aspects of cooking convenience foods extend beyond time-shifting, juggling competing commitments and scheduling strategies. At least two examples of cooking practices point towards other motivations which arise from feeling the need to reduce stress. Here, for instance, Tony (a middle-class professional in his mid-50s) argues that rather than being something potentially harmful to his health, he utilised convenience foods as a way of looking after himself

> That care for the self is, in a way, assisted by using convenience foods because I'm not putting myself under huge stress in terms of making stuff and putting all this effort in, I'm getting there quite quickly, but in a way that's satisfying to me, you know what I'm saying?

Using microwavable rice allowed him to 'sit down and relax more quickly', reducing the stress associated with his busy working life and the emotional strain caused by his recent family break-up.

In a second example, Nora (who lives in a student flat-share) talks about the stress involved in planning and cooking meals:

> I can't cook very well and I don't like it and that's why for me it's either a ready salad from Netto or scrambled eggs or frozen vegetables in a pan with cream cheese, something quick … Somehow, it simply isn't fun for me. I really don't like it, I'm already totally annoyed, when it comes to planning. If I have to bake something in three months, then it will stress me already … Simply because I don't cook, I don't have a feeling for how things go together and I don't have the patience to try it out. And when I do cook, make an effort, put something together and then it doesn't taste nice. That annoys me and then I think: 'Why bother?'

The need to reduce stress, as we can see here, can arise for different reasons. The first example shows that stress is something which is personally felt and manageable by changing practices such as cooking with new materials and meanings (turning convenience food into a 'healthier' option because it reduces stress and increases care for the self). The second example points to the importance of competences, or in this case, the perceived lack of cooking skills. Of course, regarding temporalities, the results are more or less the same—a reduced amount of time spent planning and preparing meals.

Eating

The temporality of eating a dish could again conceptually be understood as a continuum between a long sit-down meal, with plenty of time, or a very short and quick bite 'on the move' or while doing something entirely different (like working). Michael, one of our UK participants, represents both extremes to some extent. Working hard for extended periods, he tried to reduce the time needed to prepare and eat food to an absolute minimum by buying ingredients in bulk to make home-made sandwiches for lunch and stocking up on a week's supply of microwavable ready meals for dinner so as not to have to leave his flat, sometimes for days. However, mental health issues forced him to re-assess his priorities and he tried to make eating a source of pleasure again rather than a chore. His new partner also helped him to introduce a different work-life balance and increased his awareness of healthy eating/living.

These examples show that while the practices of shopping, cooking and eating all take time, their inherent temporalities can be manipulated by taking short-cuts or changing practices, including online shopping, buying in bulk, using convenience food items, and making clever use of freezers and microwaves. These creative adaptations of practices to personal and household needs point to the importance of everyday life rhythms impacting on social practices, just as Southerton (2012) says, but also to the role that dispositions and emotions play. Practices are not just changed to function in a certain way, they are also subject to constant modifications in order to make them not just practical but also appropriate (in the sense that they feel right)—in this case to reduce stress and anxieties about cooking and time spent eating.

The Temporalities of Convenience Food

As we have shown, there is an intrinsic relationship between food and time (shopping, cooking and eating all take time). This relationship varies over history and between places. In the following, and on the basis of our empirical findings, we seek to systematize this relationship further. From the literature reviewed and our own research, we posit that there are three major temporalities at work when considering convenience food and the pertinent practices of 'conveniencization' which we summarize in terms of transformations, rhythms and timings. Here, we conceptualise each of these temporalities in more detail.

Transformations

Food practices change over time and, in general, we can discern three kinds of temporal transformation: generational, life-course and lifestyle. Generational change refers to the fact that cooking and eating practices may vary historically. What, how, where and when our parents and grandparents cooked might differ in many ways from our own generation or those affecting our children. Much of this may be rooted in the historical circumstances of available foodstuffs and cooking facilities, the organisation of work, school and leisure and changing food discourses (e.g. historically contingent ideas about what is considered 'good' food, notions of thrift derived from war-time rationing etc.). Generational change can also be observed in the present, when older people prefer different types of meals than younger generations due to the tastes and routines developed over the life-course.

This brings us to the second kind of transformation over the life-course. What, when, where and how we eat changes over the life-course. It changes with age (e.g. children and teenagers are more keen on sweet and sugary foodstuffs than their older contemporaries) and the social roles one acquires over the phases of the life-course (growing up, moving in and out, becoming a parent and so on).

The third temporality is associated with lifestyle transformations. This acknowledges that temporal changes in food practices are not just due to generational transformations and changes over the life-course. Food practices can also be subject to more or less conscious individual or collective lifestyle decisions such as becoming a vegetarian or vegan, going on a diet to reduce weight or adopting new tastes and food preferences.

Rhythms

As shown in Fig. 5.1, while there are still temporal peaks in daily food consumption in the UK, eating increasingly occurs throughout the day. However, patterns within and across households are no longer defined by the workday and the time needed to cook from scratch, they appear less standardised and differences between households are likely to be more pronounced. The ubiquity of convenience food and the opportunities of 'conveniencization' (e.g. through the invention of fridges and freezers) have created ample possibilities for households to establish unique daily rhythms, tailored towards the individual schedules of household members. In our own study, we encountered a range of rhythms from highly individualised shopping, preparing and eating practices to near-standardised routines for each meal.

In general, we find that rhythms belong to different temporal cycles. Daily rhythms include the times, quantities and sequences of meals during each day. Weekly rhythms include differences in food practices in relation to particular days of the week (e.g. different foods and routines for work-days and weekends). Seasonal rhythms include the use of different food items depending on their availability and the weather (e.g. strawberries and asparagus in the spring and summer; more 'hot' meals and root vegetables in the winter and so on). Other annual rhythms include festivities and holidays (e.g. Christmas dinners or Easter breakfasts). Alan Warde (2016) even adds life-time rhythms such as special meals at weddings or funerals.

Timings

In keeping with our 'theories of practice' approach, we contend that changes and rhythms are the result of social practices to which people are recruited and which they may later relinquish. Social practices are related to material and social arrangements and historical changes over time. However, practices are to varying degrees dependent on other elements. Acknowledging the constraints that time and material resources put on food practices, our participants exercise a degree of agency by stressing the ways in which they do not just reiterate pre-structured patterns of food consumption but improvise, seeking to steer, mend or modify food rhythms and changes. Here, we refer to timings as strategic practices—similar to Warde's (1999) 'scheduling strategies' discussed above—that are directed at temporal rhythms and changes. More correctly, timing practices are about the ways in which time for food is made from the endless flow of non-food related activities and rest, and remade to fit into other rhythms of everyday life. Above, we explored a number of timing strategies which enable a deeper understanding of the temporalities of convenience food and 'conveniencization'.

The Time of Convenience Food

As we have shown, in regard to time, food can be considered convenient in a number of ways. Food may be convenient because it does not require much time to prepare. Food can be convenient because it is durable and easy to store. It can also be convenient in terms of accommodating different schedules within the same household. Many of the associated practices correspond to what Warde (1999) explored as time-saving and time-shifting. In turn, however, specific times are reserved for convenience foods. There are time zones for treats and rewards (e.g. after a long working day) or emergencies (e.g. when there is no time to shop for unexpected guests). The latter include virtual emergencies that may never happen (e.g. back-up food that is actually not bought for consumption but as an insurance against shortages).

Within the multiple temporalities we have explored in this chapter, convenience food and conveniencization seem to make sense to practitioners in regard to their busy household schedules, the preferences of household members, food security concerns, variable tastes and the logic of treats and rewards. Taking a step back and examining the longer historical implications, we may wonder if convenience food would have always made sense and be welcomed into the organisation of daily food practices or if it is a sign or symptom of modernity at large.

Sociologist Hartmut Rosa (2013) has argued that social life in Western societies has in general accelerated. While the acceleration might be welcomed in some arenas (such as medical treatment or personal mobility), it often has the effect of eroding stretches of stress-free time spent meaningfully alone or in company. Rosa contends that acceleration alone does not make a difference. If we do things faster, we have more time for other things. The specific modern form of acceleration comes from a coupling of doing more things (escalation) within shorter amounts of time (acceleration). Rosa (2013, p. 71) discerns three dimensions of social acceleration: technological acceleration, acceleration of social change, and acceleration of the pace of life. Taken together, they provide a situation where individuals and societies experience time scarcity and feel constantly pressed for time.

In the case of convenience food and conveniencization, we may consider technological innovations in the domains of food production, packaging, cooling and heating as examples of technological acceleration. Changes in dietary norms and available foodstuffs, including the proliferation of snacking opportunities, have possibly accelerated too. The main issue, however, is the perceived acceleration of the pace of life. The more activities are put into one day, the less time there is for eating (let alone shopping and cooking) or for just eating without multi-tasking. Eating under the conditions of accelerated modernity involves eating more frequently ('snacking' and 'grazing') while doing other things (on the way to work, at work, while watching TV and so on). What probably makes things worse is the de-routinization of rhythms we refer to above (cf. Shove 2003; Warde 1999). Rather than having regular times for food, food has to be 'fitted in' around the busy schedules of individuals and households.

However, our data do not fully support the hypothesis of an intrinsic relationship between acceleration, escalation and de-routinization. Seen from a different angle, integrating convenience food into domestic practices is based on routine criteria such as long work-days or occasions for treats. Cooking with convenience food allows us to have a hot meal after work at home and to go out in the evening. Speeding up the process of cooking may also mean that some families have more time to eat together. Stocks of food help to reduce the time worrying about when to buy food.

It is difficult to assess claims that people (no longer) 'have the time'. While we are certainly not implying that our respondents are less time-pressed than they express, we need to be careful assessing what practices are prioritized over the course of days, weeks and years. As practice theorists have pointed out (Shove et al. 2012; Southerton 2012), practices take time. With different practices competing for time, priorities emerge. Spending time on food practices in general seems to be in decline (as time-use data suggest). However, other practices seem to have been less affected by reduced amounts of time allocated. The question, then, is not just about how convenience food helps to save and shift time. Additionally, we might ask what are the more 'dominant' practices that are allocated sufficient amounts of time, and why in particular food practices seem to experience significant time losses. On the basis of our current knowledge, we are not able to give a full answer here, but it probably has as much to do with the acceleration of the pace of life as with the increasing availability and affordability of other options on which to spend time (like playing sports, watching TV, going out, using the Internet or meeting friends). All of these options are, of course, constrained by available resources and are subject to significant inequalities across individuals and social groups.

Conclusion

This chapter has demonstrated the multiple temporalities that are relevant to the analysis of convenience food and the process of conveniencization. It began with Shove et al.'s (2009) argument that time is about coordination and rhythm, but also involves material, emotional, moral and political dimensions. These temporalities ranged from the longer time-scales associated with historical and generational change, to the shorter time-scales of changes over the life-course and even shorter time-scales such as those associated with treats and rewards. Some of these temporalities are quite abstract and intangible such as the associations of convenience food with modernity and the multiple ways that food forms a focus for memory, while others are very practical and immediate such as the challenges of scheduling busy domestic and working lives. Having explored the routinization of shopping, cooking and eating, the chapter provides a more systematic analysis of three specific temporalities described in terms of transformations, rhythms and timings, focusing primarily on the analysis of ready meals in the UK and Germany.

Throughout the chapter, we have shown how materials such as foodstuffs and technological appliances shape the temporalities associated with convenience food, alongside political and moral arguments about time-saving and time-shifting, and the emotions that compel people to allocate more or less time to shopping, cooking and eating. As geographical scholars such as Doreen Massey (2005) have insisted, the analysis of time cannot be divorced from parallel arguments about space, such that space-time cannot be analysed as two separate dimensions. The next chapter therefore applies a similar analysis to the spatialization of convenience food, paralleling the current chapter's analysis of its complex temporalities.

Notes

1. The debate about 'labour-saving' technologies creating 'more work for mother' can be traced in Strasser (1982) and Cowan (1983). Adrian Forty (1986) describes how ideas of hygiene, cleanliness and efficiency have shaped domestic technologies since the 1750s.
2. We recognise that the naming of these meals as 'breakfast, lunch and dinner' is somewhat arbitrary and that there are significant regional and class differences in whether the mid-day meal is called lunch or dinner and whether the evening meal is known as dinner, tea or supper (among many similar variations). Referring to a warm or hot meal is also subject to significant cultural variations.
3. For more discussion of the materialization of memory through food, see Meah (2017). For related work on kitchens as sites of memory, see Meah (2016).
4. Meah and Jackson (2017) discuss the case of Hannah, a 35-year-old mother of two small children who spoke eloquently about the impact of

children on her cooking and eating practices ('Babies dominate your whole world'). When first interviewed, Hannah relied on food that could be put straight in the oven or cooked in the microwave including frozen 'rainbow' vegetables, fish fingers and other convenience foods that she could prepare while breast-feeding or keeping a close eye on her children. Revisiting Hannah eight months later, her older daughter had started school and her younger daughter (now almost two years old) was no longer so demanding. Reflecting on these changes as she prepared a 'home-made' pizza using a number of pre-prepared ingredients for the children's evening meal, Hannah concluded that cooking was 'much easier now' (ibid., pp. 2074–2075).

5. Val and Ken also spoke about the restricted seasonal availability of certain foods and the comparison between tinned and fresh fruit ('you can't have fresh peaches [in March]'). This is another kind of temporality associated with convenience food, discussed in more detail in Chap. 9 on health and sustainability.

6

The Spatialities of Convenience Food

Introduction

This chapter aims to shed light on the spatial dimensions of convenience food in order to better understand the contexts in which the spatially dispersed practices of 'conveniencization' literally take place. The development of convenience food is often associated with retail distribution, food preparation and eating taking up ever wider spaces as supply chains have becoming increasingly globalized and food consumers have become increasingly disconnected from places of agricultural production. Convenience food is one of the classic examples of increasing 'food miles', a concept that refers to the distance travelled by foodstuffs and packaging material before they reach the actual point of purchase.[1] The preparation of meals has also become distributed across space and time with the advent of convenience food or ready-to-use ingredients, and eating is now more spatially diffuse as modern take-away food, snacks, sweets and sandwiches can be consumed virtually anywhere.

The spatial widening of food practices (including acquisition, preparation, eating and disposal) has been accompanied by increased social anxieties (Jackson 2015). 'Food miles' have been debated as a sustainability

P. Jackson et al., *Reframing Convenience Food*,
https://doi.org/10.1007/978-3-319-78151-8_6

issue. Factory-based on-site meal preparation has raised concerns about food quality, health and hygiene (including concerns about vitamin loss through multiple cycles of heating and cooling and their associated energy costs). Eating 'on the move' and in public places has been criticised as an uncivilized way of eating, seen as a key driver for the erosion of communal eating and linked with the demise of the family meal, giving rise to multiple social concerns. Taken together, the widening spaces of food production and consumption have led to a greater conveniencization of food, but they have also become a source of widespread concern, encapsulated in disparaging references to the 'McDonaldization of society' (Ritzer 1993).

As part of our wider commitment to reframing the subject, this chapter will challenge these commonly-held assumptions about the spatialities of convenience food. We argue that these spatialities are complex and that the implications for social life, health and sustainability are at least ambivalent. The argument begins with a brief overview of the concept of spatiality before drawing on a range of examples from our four cases to analyse the spatialities of food distribution, preparation and eating. The chapter concludes by emphasizing that the normative implications of food's diverse spatialities are far less straightforward than the widespread anxieties regarding convenience food suggest.

Spatialities

Convenience foods involve a series of spatial phenomena. For the sake of conceptual clarity and for analytical reasons, we theorize spatialities in parallel with our discussion of the temporalities of convenience food in the previous chapter. There, we introduced Southerton's (2003) three dimensions of time related to social practices: time as a resource, the time that practices demand, and temporal rhythms. Similarly, we contend, the spatialities of social practice are best understood in terms of space as a resource, the spatial demands of practices, and the spatialities of everyday life.

Space as a resource refers to the fact that the globe is spatially finite and that, if space is taken up for one thing, it is often not possible to use it for another. Space is also a resource in the sense that it provides or limits the spatial expanse needed by individuals and social groups to pursue certain activities and lifestyles including the production, consumption and discarding of things such as foodstuffs. There is also an implication in this argument about resource use that space is not only used but can be used up. This is characteristic of the negative associations that are often attached to the idea of 'consumption'. However, a more relational view of space focuses on the interrelated material entities that can be used, used up or altered in the pursuit of social practices where space is not a container or fixed resource but is actively produced and reproduced through multiple (sometimes competing) uses.[2]

That *practices demand space* follows from one of their constituent elements: practices are always bodily activities (Reckwitz 2002). Bodies obviously take up space and if a practice is to be carried out, the body that performs the practice takes up space while doing so. Practices thereby vary in the spatial expanse necessary for their performance. Producing coal for energy consumption, for example, is a social practice that requires access to spatially extensive coalfields. The practice of eating, however, can take place on a very small scale. In short, practices prescribe the minimum amount of space needed to conduct the activities involved.

The *spatialities of everyday life*, in turn, counterbalance the spatial demand of practices. Societies and individuals can draw on their own characteristic mobilities and set of places. In technological and capitalist societies, the spatialities of everyday life vary from place to place and from person to person. Someone living on a low income has more limited mobility and access to other places than someone who is better off. Likewise, living in the countryside can have a significant impact on geographical mobility and access to places compared to living in a city. Practice theorists refer to personal and collective 'place-path arrays' (Schatzki 2009) which may also differ by demographic characteristics such as age, gender and ethnicity or by occupation and preferred leisure-time pursuits.

Spatialities and Convenience Food

The conceptual approach to the spatialities inherent in social practices outlined above can now be more clearly defined in relation to convenience food and the practices of conveniencization. There are, again, at least three dimensions to how convenient food practices are spatially entangled: spatial relations, place, and spatial change.

The *spatial relations* of convenience food and convenient food practices are made up of the spatial extensions and networks necessary to enact or sustain a certain practice. For example, cooling and heating convenience food products require energy which is mostly produced elsewhere. It relies on a network of energy plants, cables, houses, sockets and so on. The same is true for water which is needed for many cooking practices and where we rely on it 'conveniently' coming out of the tap whenever needed. In general, convenient food practices depend on the ability to draw things in from elsewhere, thus producing a dynamic network of things on the move and in circulation.

Place is needed to enact a certain practice. Cooking, for example, needs a place that makes cooking possible. Technically, cooking can be done wherever it is possible to produce heat. Nowadays, however, cooking is mostly done in kitchens. What and how much can be cooked depends on the size and infrastructure of the kitchen. In addition to the idea of place as a sort of resource, our 'sense of place' (Tuan 1977) produces collective and personal dispositions toward the appropriateness of certain practices in certain places and the persons who are deemed to be the acceptable practitioners in these places (cf. Cresswell 2004). Where we eat, for example, depends as much on personal preferences as on socially shared (even if contested) norms about the appropriate places for specific forms of eating for specific social groups (cf. Bell and Valentine 1997).

Spatial change occurs throughout the aforementioned spatial practices. Changes can be intentional or unintentional. Buying a microwave, for example, changes the infrastructure of the kitchen and can make it a better place for convenient cooking practices. The use of energy, on the other hand, transforms our planetary resources, increasing greenhouse gas emissions and atmospheric pollution, for example. In a way, spatial imaginations are also part of the transformations of space and place. Ideas about the perfect kitchen, where food comes from, or how far it is to the nearest supermarket all impinge on the ways we deliberately and unconsciously manipulate and transform space and place. Often, our spatial imaginaries bear little relationship to the actual conditions in which food is produced as is apparent when one considers ideas of authenticity or constructions of exotic food, for example (cf. Heldke 2003).

This chapter draws empirically on all four of our cases, giving priority to our study of workplace canteens. Canteens are hubs where cooking, eating and work all come together in 'convenient' ways. The spatial intensity of the canteen makes it a formidable study site which helps us understand the multiple spatialities at work in the process of conveniencization. After a brief history of canteens as institutionalized sites of collective food provisioning, the following sections analyse the spatialities of different kinds of convenience food and the spaces associated with the conveniencization of food. We look in some detail at the practices and spatialities of networks and circulations, food preparation, and eating, clarifying the spatial relations, places and spatial changes involved in the production and consumption of convenience (and convenient) food.

Canteens as Institutionalized Spaces of Collective Food Provisioning

The history of canteens as institutionalized sites of collective food provisioning demonstrates how the conveniencization of workplace eating was always part of their raison d'etre. According to Thoms, the German word *Kantine* is of military origin:

> Originally it meant an inn for soldiers in a fortress, where only alcoholic drinks were sold … Dictionaries mention *Menage* (tied accommodation) or *Speiseanstalt*, in the sense of provision of complete meals, from the 1820s, and they were clearly associated with the purpose of saving money by housekeeping together with others. The term and concept were taken over by industry from the 1850s. Consequently, the conception was one of canteen food specifically meant for young, strong and healthy men doing hard physical labour. (Thoms 2003, p. 352)

Nowadays, the clientele in canteens includes male and female employees rather than being limited to men or manual workers (Thoms 2009,

p. 142). Though the term canteen is nowadays often replaced by other terms such as cafeteria, *Kantine* is still used in everyday German language.[3]

The spatial and temporal organization of eating at work is intrinsically connected with the organization of work itself including the location of the workplace in relation to employees' homes and other eating facilities, the kind of work being undertaken, the number of workers and the length and coordination of breaks—especially around lunchtime—as well as the timing of the working day. Apart from monasteries and military establishments, the introduction of mass catering at work coincided with the industrialization of labour. The spatial separation of domestic life and paid employment was a precondition for eating outside the home (Tanner 1999, p. 210). The organization of workers in assembly lines, mills and factories also raised issues of temporal scheduling and spatial coordination which canteen meals facilitated. A disciplined industrial workforce necessitated new conceptions of time, where family time was subordinated to industrial time (Thompson 1967; Hareven 1982). As Luks (2009) argues, the organization of canteens reflects the spatial-temporal organization of Fordist work with its ideal of a smooth, efficient and dynamic workflow, providing an industrial mode of well-nourished working bodies (ibid., p. 49).

Canteens have been interpreted as places which enhanced capitalist control over the working body (Uhl 2012) but also as places of improved social welfare, ensuring the provision of food for working people while controlling their access to alcohol. As Thoms (2009) argues, 'canteens did not evolve out of collective action but were created by factory owners, their costs calculated as an investment in the creation of healthy and strong workers, their bodies interpreted as machines that needed the right fuel in sufficient quantities to work properly' (ibid., p. 131). While some employers subsidized canteens in the nineteenth century, it was not until the late 1950s that the majority of workers had a canteen lunch in the former Federal Republic of Germany (Uhl 2012). In the former German Democratic Republic, canteens were promoted by the government in order to ensure national food security and to facilitate women's participation in the labour force (ibid., p. 364), while in the FRG, canteens were seen as a 'welfare institution', opposed to the ideal of home-cooked meals prepared by the loving mother (Thelen 2006, p. 54). In each case, though in different ways, workplace canteens contributed to the conveniencization of food provisioning as well as serving a range of other social functions.

Networks and Circulations

While canteens are a good example of the spatialities involved in providing convenient workplace food, we might also probe the spatial dimensions of other forms of food supply including the networks through which food is sourced and the circulation of food through retail and distribution channels. How does food come to us? While we may not always think about the provenance of the food we eat as we go about our everyday lives, there is an increasing consciousness that foodstuffs can be sourced from virtually anywhere on the planet. Notions of seasonality and locality have been disrupted by the globalization of supply chains and their analysis reveals the increasing interconnectedness of places in terms of food production and consumption (Goodman and Watts 1997). These trends have been accompanied by increasing public anxiety about food safety and security as demonstrated by Susanne Freidberg's (2004) study of the links between the production of French beans in Burkino Faso and Zambia and their consumption in France and Britain.

However, there is a debate on how many miles food travels until it reaches the final place of consumption and whether more miles always means less sustainability (Defra 2005). Despite growing public concern about 'food miles' (as discussed in Chap. 9), there has been limited impact on foodstuffs being shipped around the globe and limited influence on shopping and eating practices. The increase in the popularity of convenience food (documented in Chap. 2) suggests that 'food miles' are of little concern when it comes to actual purchasing decisions. Food practices are socially embedded and, as a result, decision-making processes are slow to change.

Buying convenience food items, for all sorts of reasons, weakens the potential impact of the 'food miles' concept further. People may care about where an apple comes from, for example, but when buying a ready meal they are much less likely to ask where basic ingredients such as wheat, starch, sugar or eggs come from. Similarly, consumers may be concerned about the provenance and welfare of the chicken they cook for Sunday lunch but rarely seek information about the origins of the poultry used in a chicken tikka masala dish or chicken nugget meal (Jackson 2015). Sometimes, of course, our taken-for-granted practices are disrupted as occurred in the UK in 2013 when the content of beef lasagne and frozen burgers was found to have been adulterated by the undeclared inclusion of horsemeat. As the complexity and extent of meat supply chains became apparent, there was heightened concern among consumers regarding questions of transparency and regulation, and several major food retailers experienced a costly decline in consumer confidence. Even in this case, however, trust was quickly recovered and sales of meat-based convenience products were soon back at their previous levels.

When considering the conveniencization of food more broadly, it should be acknowledged that the origin of ingredients and the pathways from farm to fork can be as obscure when cooking from scratch. This is particularly so with canteen food, where diners order a meal about which they often have little or no information regarding the spatial networks, flows and circulations that are necessary to provide the food they eat. Consumers are forced to abrogate responsibility, placing their trust in the catering establishment regarding questions of food safety, provenance and quality control.

Here, for example, are some of the ways that customers in a hospital canteen, interviewed by Christine Wenzl, responded to questions about their choice of specific meals, ranging from the highly pragmatic to the slightly more principled:

Markus (who eats Currywurst):	It was acute hunger for something salty, a savoury meal.
Manfred:	Because I know this [dish]. It's not very healthy, that Currywurst, but it tastes good.
Klaus:	Actually, I came here just to get some coffee and then I could smell the sausages [laughs] and, uhm, I know it's not good for me, but the yearning, that's what did it.

Paul:	I am not the biggest meat fan … I am not at all vegetarian; I eat meat now and again but the meat here is not very high quality, I think it is safe to say. It's probably meat at a very low price. That's why I eat in principle a bit more cautiously, I'd say, and that's why I opted for the vegetarian dish, because I think, that many of the vegetarian dishes here are actually quite tasty.
Christine:	*Why did you choose this dish?.*
Lars:	Because the fish was not there.
Bernd:	Most of the times, I eat vegetarian. And this is the vegetarian dish.
Carla (who chose a baked potato):	I went for the visual appearance.
Petra:	I had the fish. Fish is always good. I don't make it at home.

This range of answers implies that the choice of a specific dish is highly embedded in the rhythms and logics of everyday life and in broader discourses of health, sustainability and 'good food'. However, the spatialities of food—in the sense of where food comes from and how and where it is produced and who prepares it—were barely touched upon. Reservations were common regarding the quantity and quality of meat, although it was less clear if those reservations originated in ethical, environmental or health-related concerns.

In general, going for lunch in a canteen is a highly 'routinized type of behaviour' (Reckwitz 2002). Although some respondents scrutinized the menus for the whole week and had a more methodical approach to their selection of dishes, many decided spontaneously. Moreover, some dishes sold out more quickly than others, limiting choice. In some canteens, where each dish has a separate queue, people opt for less popular dishes but with shorter waiting times. When faced with these constraints, one would think that frustrated canteen customers would turn to other sources for their mid-day meal. However, that did not occur to most of our participants. While we encountered many forms of place-mixing (using different places for lunch, including eating at the desk), those who went to the canteen on a regular basis often responded to the question 'why are you eating here' with the simply response: 'because I am hungry'.

It seems that the context of eating in the canteen shapes the ways in which 'food miles' and provenance are made (ir)relevant to consumers. So, even if simple ABC models of behaviour change which assume a linear relationship between attitudes, behaviour and choice (cf. Shove 2010) do at times change consumption practices through an increased awareness of ethical, health and sustainability issues, huge variations remain and must at least partially be understood through the socio-spatial context within which consumption practices take place.

Just as eating in the canteen obscures the complicated geographies of foodstuffs moving around the globe, the packaging and context of eating convenience food similarly gloss over the underlying geographies of production and trade. However, the hidden geographies of convenience food do not go entirely unchallenged (cf. Harvey 1990). While ready meals, take-away food and canteen meals provide very limited information about the previous whereabouts of the foodstuffs consumed, strategies are increasingly observable to make these connections more visible and understandable. For instance, a typical German 'Imbiss' in Bonn that serves many workers, tradesmen and craftsmen working in the area and which sells mainly meaty take-away dishes such as Bratwurst with French fries, has a note in the window explaining that virtually all its meat is obtained from a regional slaughterhouse. Of course, this does not tell customers where the animals came from, the conditions in which they were reared or what the standards of the slaughterhouse are. However, it is interesting that even a low-key food site like Imbiss nowadays feels compelled to inform customers about—at least a part of—the geographies of its products.

Our case study of meal-box schemes in Denmark provides another example of people actively seeking to influence parts of the food supply chain. Meal-box schemes provide customers with an alternative way to obtain food in a very convenient manner and to get to know more about the pathway of their food—at least in terms of delivery. Similarly, in Sweden, the mothers we interviewed were concerned about the health and sustainability of the food they served to their babies, including its organic credentials, though they were less concerned, in general, about its physical provenance.

Food Preparation

The stages of preparing convenience food can be thought of as a form of cooking practice which is dispersed and distributed across time and space. Not only does convenience food and the technical appliances for cooling and heating help to disperse the cooking process through time, but also in space. Ready meals, for example, have been pre-prepared to a significant degree, in some cases only requiring heating. This part of the cooking process happens outside of domestic homes and kitchens, taking place in large-scale food factories for example. Likewise, cooking in canteens is a relatively dispersed activity. Some of the foodstuffs come in as much pre-prepared as more conventional convenience foods, ready to consume in private households. Other foodstuffs are prepared from scratch on-site, but also shifted around a lot more than in the small kitchen of the average household.

In the canteen kitchens we visited, staff claimed that virtually all of their meals are prepared from scratch (including mashed potatoes, gravies and broth). On the other hand, all the chefs and canteen managers we interviewed had some experience of using convenience foods in their workplaces:

> Especially in the catering business and if that is the business you're in … you're sort of, I'd say like propaganda, informed by your suppliers. And they have new products and tell you 'that's what we have new' and 'try it'; suddenly there is this meatball, before, you had minced meat and this is what you had to make a dish, now there is this ready-made meatball. Alright, we test it—good. And then you would find certainly in one way or another more products, where you say, well, that's something I could use. (Head chef of training centre canteen, male)
>
> And look what's in there that doesn't belong in there, things you wouldn't put in yourself… The last time my manager brought some convenience…, some marinated meat [from the local butcher], I asked what's in there because I have customers who really do have allergies. Well, and at first they said, it's free of anything, but then I did look, and there was a label detailing the ingredients, and it included glutamate and mixed herbs … well, that is not acceptable for me, even if it is cheaper. Because then I need to explain again to my customers that there is something in there. (Chef of court canteen, female)

> We work very little with convenience products, sometimes, of course. But we have made it our motto, as long as we have enough staff—we have all learned our trade properly and first-hand—we will as far as possible abstain from the use of convenience products. For one thing it has incredibly many additives, allergens—who knows what's in these convenience products. For another thing, it has been proven right, if I cook from scratch, I cook cheaper. (Chef of court canteen, female)
>
> In general, the highest cost in a catering business is staff, and through convenience, the use of convenience products, perhaps I can reduce the number of staff. These are reasons for using convenience products—every business needs to find out by themselves if they agree to the use of convenience products or not. (Head chef of training centre canteen, male)

In many cases, a trade-off seems to be considered. Using convenience food saves time during the final cooking process. Thus fewer staff are needed in the kitchen. On the other hand, all respondents claimed that convenience foods are more expensive and that they can cook more cheaply by using fresh ingredients. In addition, chefs claim that meals cooked from scratch have a better taste and will be valued more by their customers.

This emphasis on cooking from scratch has implications for the spatialities of the cooking process. As in every standard European kitchen, we expect to find the usual appliances but on a larger scale. This is, in general, true of all canteens. However, there are a lot more machines like blast chillers, convection ovens or conveyor dishwashers. The cooking is done by several persons and divided into set tasks. In some canteens, unlike the ones we visited, more machines and the use of convenience products may actually reduce the size of the kitchen and the number of staff. Some machines do important cooking tasks such as snap freezing and reheating.

The spatiality of the kitchen is not only influenced by the size of the rooms, the machines and the staff. In particular, health and safety regulations have a visible impact on the spatial ordering of canteen kitchens. Unlike private households, where everyone decides more or less on their own what parts of the kitchen are considered clean and which are dirty, national and EU regulations have divided up canteen kitchens into clean and dirty spaces, in order to avoid cross-contamination. Staff who work a shift 'on the dirty side' are not allowed to work that day on the 'clean side'.

Regulations also have spatial effects beyond the kitchen. Within the EU, regulations require all food serving businesses to inform their customers about allergenic ingredients. This has led to changes in what food canteens buy and how meals are prepared. In our sample, the evidence suggests that more canteen chefs are now cooking from scratch to make absolutely sure they know what is in their dishes. However, there are also examples where the opposite happened and more convenience foods are used because allergens are listed on the labels.

Our kitchen staff respondents also complained about the effects of allergen labelling on their cooking practices and the ways in which their food is perceived. Naming all allergens in every meal means that the ingredients for each dish have to be carefully planned well in advance. Making last-minute changes or spicing things up a little is no longer possible. As one of our respondents phrased it, in a somewhat derogatory and offensive manner: 'This way, I don't need experienced chefs anymore, even a housewife can do the cooking'.

According to another chef, the way that allergens are made transparent has a negative impact on how people perceive food in general.

> Roux, for example, and the like, it has nothing to do with chemicals or toxic or so, but now there is a number and everyone thinks, 'Well, look what's in here'. I do not like this. I think it would have been better that people with allergies come to us and say, like it has been before, I have a nut allergy, are there any nuts in there, or onion allergy [...] now everyone is alarmed, thinks they are being poisoned.

Eating Spaces

As mentioned at the beginning of the chapter, eating can nowadays occur at virtually any place and any time. The meal situation has certainly become less and less dependent on set times and places (Hirschfelder 2001). Nevertheless, people do eat at certain places and times even if these are less standardised than before. And while there is increasing evidence that people eat at home in more and more places besides the kitchen or living room table, such as couches, bedrooms, floors, desks etc., and at times they deem convenient, there is also evidence that people are eating less and less at home (Warde and Martens 2000; Warde et al. 2007). Within food research, the classic distinction is between 'eating in' (at home) and 'eating out'. However, while eating out is commonly conceived of as eating a meal in a restaurant or similar establishment, meals consumed in school, at work or near to work, in canteens, hospitals, elderly homes and so on hardly figures in the research on eating out. This is probably because many of these eating places do not neatly fit into the dichotomy of eating in/out and blur that established boundary.

Boundaries between eating in and eating out are blurring in other ways too. British supermarkets originally developed ready meals to cater for customers' changing tastes, following holidays abroad or increased experience of eating out, providing restaurant-quality food at home (Warde 1999). Today, supermarkets are offering customers the opportunity to 'dine in at home' with food that provides a cheaper alternative to eating out. So, for example, Marks & Spencer offer a main course, side dish and dessert for two people with a bottle of wine for £10, designed to rival restaurant meals taken outside the home (http://www.marksandspencer.com/c/food-to-order/dine-in, accessed 8 December 2017).

Like restaurants, workplace canteens are different from eating at home. However, the frequency and recurrence with which they are used, gives workplace canteens some properties that are quite different from restaurants, take-aways or other options for eating out. For example, more home-like spatialities such as familiar seating patterns can be observed in many canteens and catering staff get to know the food preferences of regular customers, interacting with them socially on a more frequent basis than in most restaurants.

An intriguing example comes from a male respondent who told us how he—unknowingly—disregarded the socio-spatial order of a courthouse canteen (Amtsgericht) when he was a newcomer to the place. Our respondent had arrived quite early and seated himself on the biggest table. After a while, when new customers arrived, he wondered why some of them frowned upon him and no one came to join him on the large table but instead huddled together at much smaller tables. Only later did he discover that he had unwittingly placed himself at the table that was usually occupied solely by the judges, a convention that was observed in practice without the need for official designation. In other places, such as hospital canteens, seating orders are more explicit with reserved tables for staff.

Canteens can also be seen as a convenient place for meeting people and socializing. In a sense, they make up for time not spent eating in company at home. This is obviously very dependent on the personal and social context. While in the UK canteens are often considered a less desirable place to eat during the workday (where a sandwich lunch at the office desk is more common), in other European countries canteens are considered an obvious choice for lunch (and sometimes even for breakfast and dinner). Estimates range from 8 to 10 million meals served in German canteens every day (Max Rubner-Insitut 2008; Statista 2017).

As far as we can tell from our research, the practice of eating in canteens involves several ways of socializing and various forms of social interaction. Of interest here are the three different but often interrelated forms of enjoying a break from work, doing more work by using the collective meal as a convenient time for conferencing and debriefing, and using the meal as an opportunity to reflect on work:

> I try to use this as a break, to decompress. But very often, this decompressing happens through talking stuff through that somehow has happened.
>
> Well, what do we talk about? Certainly we do exchange a few private things. I mean, some of us, we do know each other for quite a few years now.
>
> Sometimes I think I can use a little quiet when the morning has been very demanding, then I do not necessarily have to have a chat. Also I just sit there and listen.
>
> A little bit of everything: sports, job. Well, men talk a lot about football; work sometimes.
>
> We have people from other departments, who sit with us…, they talk through with us things that the bosses came up with and we, through talking, put the whole thing into practice.
>
> I'd say we do speak about work initially, but move on to more private matters. It's a mix. But often an opportunity to talk to the colleagues from other departments … organizational stuff.

> The working group usually has lunch together … And I know that I always meet many colleagues here. We have had professionally good councils and here you can talk something over quickly.

These examples include a number of ways in which the time and the place of the canteen are used. It ranges from enjoying a break alone, quietly in company, or decompressing and talking about private matters. There are also examples of conferencing and debriefing about work issues, all of which are considered 'convenient' ways of eating. Finally, as we know from our observations, canteens provide a space to reflect on work, including making jokes, letting off steam or boasting in front of colleagues. Just as Waddington described in the case of the British police, the canteen also provides a space that helps to cope with stress and other problems (not necessarily work-related), where 'canteen culture' plays a key role in shaping the wider occupational culture:

> [T]he canteen offers one of the rare opportunities for officers, whose actions on the street are normally 'invisible', to engage in displays before their colleagues. Here officers retail versions of events that affirm their worldview: the canteen is the 'repair shop' of policing and jokes, banter and anecdotes the tools. (Waddington 1999, p. 295)

For one British participant, the lunch period was regarded as an opportunity to take a 'tactical break' from work and to enjoy protected time in the middle of the day. It was a conscious practice of switching off. Here, Tony explains how this might be a meal enjoyed at a venue on site, or away from his workplace altogether:

> I always have lunch break, I take time for lunch every day, have lunch, it might just be in the café [in his office building] or I might go somewhere off [site], but I do actually like the idea of … I just wouldn't sit in my office with my sandwiches, so I do make a point of taking myself out
> *Angela: That's quite a mindful, a kind of conscious thing?*
> It's very conscious, yes, in my diary it's blocked out: 1–2. I'm not available. I'm flexible obviously [but] for me, it's quite important to have that, just break, basically.

Another British participant, Maryam, reported how her husband, Imran—who worked for the same organization as Tony—enjoyed eating both breakfast and lunch in either a workplace canteen or café:

> My husband always has his breakfast and his lunch from work because he finds it more, like he go there, because they got a canteen or a café, he get the breakfast, you know, he like the sultana bread, teacake, he just love the teacake and a cup of tea, that's his breakfast. He want it from there because it's hot and he likes to sit down and read a paper or whatever.

Imran enjoys taking his breakfast away from his wife's early morning 'battle' with their teenage children, trying to persuade them to eat healthily. The canteen provides a quiet place to enjoy a drink and something to eat and an opportunity to read the newspaper—activities which revolve around neither family nor work.

While the spatial dimensions of the canteen may appear unique and possibly unrelated to other forms of conveniencization such as ready meals, it is remarkable how conveniencization in general has an effect on routine spatial orderings and social interactions. As we have seen in the previous chapter on temporalities, convenience food can sometimes be used to free up time for other family activities. In some cases, customers who eat a hot lunchtime meal in the canteen will eat a simpler evening meal, the practices in the workplace having a spill-over effect on consumption spaces at home. In other cases, canteen customers will take food home to eat in the evening when they do not have the time or inclination to cook. In general, convenience food has made the geographies of eating more fluid and diverse.

There are many other examples in our empirical material where convenience food enhances participants' mobility, altering the spatiality of their everyday lives. In Britain, for example, Maryam's children eat biscuits and cake-bars on the way to school, while her son stops at the chip shop on the way home from school. Another participant, who works as a postman, eats scones and tea-cakes in his break at work or consumes breakfast-bars while doing his rounds as he leaves home early in the morning without time to eat a 'proper' breakfast. Similarly, in Sweden, jars of baby food enable mothers to increase their mobility, allowing them to eat 'on the move' and to respond to the baby's urgent demand for food when other options are not available. As these and other cases suggest, the spatialities of convenience food are fluid and multiple, enabling and constraining a range of eating and related social practices.

Conclusion

As the examples in this chapter show, the spatialities of convenience food (such as eating a hot meal in a canteen) are constituted through at least three interrelated spatialities. First, convenience food relies on resources drawn in from elsewhere. Food is virtually never consumed exactly where it is produced. However, 'conveniencized' food often has a complex geography in terms of the places where it was pre-prepared before being heated up and served as a meal.

Second, the networks and circuits of foodstuffs are anchored in specific places where meals are prepared and consumed. Kitchens play a role with their technological capacities and the competent bodies of those who cook. Where people eat is also important. The places where meals are consumed can lend themselves to many different purposes. In the case of workplace canteens, we have seen how the place can be an extension of the workplace as much as a place for private communication, decompressing or simply taking a break.

Third, spatial change occurs through changes in and through practices and material arrangements. What foodstuffs and degrees of convenience chefs and private cooks choose has a direct bearing on where food comes from and to what degree it is pre-prepared. Many canteen managers were reluctant to use convenience products, unwilling to surrender responsibility for health, safety and hygiene. However, the introduction of rules and regulations such as the allergen list has brought about changes both to kitchen practices and to the spaces of eating.

Finally, our research suggests that there is no straightforward connection between convenience food or conveniencization and its impacts on sustainability, health and social life. The geographies of convenience food may be more complex in the sense that more places are involved and there is evidence that modern foodways are space and resource intensive. But some conventional foods, using fresh ingredients cooked from scratch, may be equally or less transparent in terms of their environmental impact, geographical provenance or health implications, their lack of labelling in terms of origin or ingredients making them more opaque. The spatialities of convenience food are manifold and complex, some being more spatially intensive and others more extensive than others. The consequences of these inferences in terms of health and sustainability are explored in more detail in Chap. 9.

Notes

1. The concept of 'food miles' was introduced by City University food policy professor Tim Lang ('The man who invented food miles', *The Observer*, 25 March 2007). While the concept has served as an effective way of encouraging public engagement with the environmental issues associated with food, it has been criticised as an analytical concept as discussed in more detail in Chap. 9.
2. The production of space was a key concern of the French social theorist Henri Lefebvre (1991).
3. In English, the word 'canteen' still has a dual meaning, as a workplace restaurant and as a bottle or container for food and drink used by soldiers, walkers and campers.

7

The Moralization of Convenience Food

Introduction

According to Rozin, 'Moralization is the process through which preferences are converted into moral values, both in individual lives and at the level of culture. Moralization is often linked to health concerns … It is significant because moralized entities are more likely to receive attention from governments and institutions, to encourage supportive scientific research, to license censure, to become internalised … and, in at least some cases, to recruit the emotion of disgust' (Rozin 1999, p. 218). This chapter explores the moralization of convenience food, with particular reference to the consumption of ready meals.

Writing twenty years ago, sociologist Gerhard Schulze (1997, p. 38) suggested that the rhetoric of everyday verbal intercourse was undergoing a significant loss of normative concepts, with notions such as 'sin', 'guilt', 'good', 'evil', 'decency' and 'selfishness' dropping out of collective memory. His prediction that this would not, however, presuppose the end of the history of morality has proven well-founded: fast forward a decade to the damning representation of a group of British mothers as 'sinner ladies' in the country's tabloid press. Their crime? To be caught passing children

P. Jackson et al., *Reframing Convenience Food*,
https://doi.org/10.1007/978-3-319-78151-8_7

take-away food through the railings of Rawmarsh school in South Yorkshire which had undergone a culinary make-over via television chef, Jamie Oliver's, school dinners campaign.[1] The resulting furore which reverberated across the world's media (Fox and Smith 2011) points toward a seemingly indissoluble relationship between food and morality, as well as reinforcing the applicability of Rozin's observations—quoted above—as a pertinent framework for understanding how moralization occurs, in what contexts and with what potential outcomes.

Food provokes a moral response in many different circumstances. For example, the alleged decline of the Sunday lunch has been equated to a full-blown 'moral panic' in the UK where *The Independent on Sunday* newspaper ran a campaign with restaurateurs and celebrity chefs describing Sunday lunch as 'sacred' and its potential loss as 'sacrilegious' (Jackson et al. 2009). Nor are these issues limited to the UK. Writing about changing meal practices in suburban Norway, for example, Bugge and Almås (2006, p. 210) argue that 'To cook a proper dinner for one's family is an important part of a woman's understanding of her own identity and an implicit part of realizing the ideal family and the ideal home'.

Concerns regarding the proliferation of diet-related illnesses including coronary heart disease and Type 2 diabetes, along with fears of an obesity 'epidemic', have contributed to an increasingly vocal, scientifically-based secular morality (Katz, cited in Rozin 1999, p. 220) which has been mobilized in what Haggerty and Ericson (discussed in Rich 2011) describe as a 'surveillant assemblage' of socially and culturally approved eating and feeding practices. Today, morality is—to a large extent—linked to scientific claims and notions like 'risk', 'safety' and 'choice' (Montelius and Nygren 2014) rather than 'sin', 'good' or 'evil' (see also Hunt 2003). Indeed, decisions about everyday food consumption including breast- or formula-feeding, infant weaning practices, what is bought (and from where), how it is cooked, stored and disposed of, where it is eaten (and with whom), are all accompanied by a relentless moral refrain to do more of one thing and/or less of another.

While all food inhabits a moral space (Lupton 1994, p. 681), it is perhaps unsurprising that 'convenience foods' have become a target for apportioning blame regarding current concerns about diet-related ill health since many ready-to-cook or ready-to-eat products have not been renowned for their healthfulness. Instead, their use has come to epitomise much that is deemed 'bad' about the contemporary social order. Critics associate convenience food with a lack of knowledge about foods and their properties, an absence of the skills required to cook and care for oneself or one's family, and an inability to transfer such knowledge to future generations. Indeed, the use of convenience foods has become a synonym for bad parenting and unhealthy lifestyles (Jackson and Viehoff 2016). Given the characterization of the Rawmarsh mothers as 'junk food mums' in the British press and on social media (Fox and Smith 2011), and Jamie Oliver's various vilifications of the parents who feature in his programmes,[2] we might be forgiven for assuming that the moralization of food consumption is peculiar to the current period. While the role of the media and the impact of social media in particular, has undoubtedly extended the ways in which discourses concerning healthy bodies, 'good' practices and responsible consumers are circulated, these discursive traditions are not restricted to the contemporary period.

This chapter examines the moralization of convenience food, frequently articulated in terms of the contrast between convenience and care (Warde 1997). Challenging this opposition, we argue that the use of convenience food can be justified as an expression of care for family members as well as distant others. We also problematize the concept of 'food choice' suggesting that the term is often used ideologically rather than as an analytical category (cf. Jackson 2016). In what follows, we trace the moralization of food, highlighting how present concerns have their parallels during periods characterized either by the absence of the 'choices' we currently take for granted, or the technological innovations—in product development, storage and transportation—upon which such choices depend.

A Brief Genealogy of the Moralization of Food

As public health scholar John Coveney (2000) has discussed in detail, the relationship between food and the body is a problem that has preoccupied Western thought since antiquity. Drawing on the work of Foucault, and in particular his discussion of governmentality, Coveney points out that the ethics of our relationship with food 'determine how the individual is supposed to constitute himself as a moral subject of his own actions' (On the genealogy of ethics, cited in Coveney 2000, p. xii) and that these relationships are socially and historically patterned rather than being conceived by the individual alone. These observations are borne out if we examine the ways in which ethical considerations were understood in antiquity as enabling individuals to be proper citizens. For example, the ancient Greeks—who did not conceptualize pleasure (be that in relation to sex or eating) in the problematic way that characterized subsequent periods—believed that pleasure could be attained from asceticism, moderation and self-mastery, by living modestly and limiting one's desires. Indeed, self-control was believed to develop the capacity for reason (Coveney 2000, p. 33). Among Roman citizens, pleasure was similarly understood as involving self-mastery and, in this context, food was regarded as playing an important role in facilitating care of the self. Here, self-mastery was often seen as preparation for some kind of suffering and a demonstration that one could train oneself to withstand particular privations and practise self-control (ibid., p. 36). In antiquity, the pleasure associated with food was balanced with restraint, enabling individuals to be good citizens.

The moral dimensions of food did not start to become problematic until the Christian period, when food and 'appetite' became more overtly linked with lust and the pleasures of the flesh, while illness and disease were considered to be manifestations of evil and sin. Where fasting had previously been associated with self-mastery and care, indivisible from pleasure, among early Christians it became a vehicle for public confession and a display of penitence (Coveney 2000, p. 38). Importantly—in terms of contemporary framings of food—it was also recognized that since food was exterior, it was also polluting to the body (ibid., p. 48). Thus, while conduct concerning one's relationship with food was conceptualized as a matter of personal ethics during antiquity, among Christians, eating became a way of expressing one's duty to God through the correct use of diet (ibid., p. 54). By the turn of the sixteenth and seventeenth centuries, 'civilising of the appetite' (Mennell 1987) had become a project which extended beyond the religious classes, involving a code of conduct for a now increasingly interdependent wider population. Drawing on the work of Elias and Foucault's arguments concerning governmentality, Coveney (2000, pp. 47–48) interprets these developments as reflective of 'the emergence of a mundane administration of citizens in terms of the collective welfare of the state … the beginnings of a calculated and state authorised conduct around food'. Specifically concerned with the genealogy of discourses about nutrition, Coveney's examination reveals that while contemporary concerns with 'food choice' are seemingly inseparable from discourses of morality, the modern science of nutrition that emerged in the twentieth century tapped into a much deeper tradition in Western thought that problematized the relationship between food and pleasure (ibid., p. xii).[3]

The history of dietary advice in the US has been charted by Melanie Du Puis in her recent book on *Dangerous Digestion* (2015) where she shows how the ingestion of food has long functioned as a metaphor for wider concerns about social control. From America's founders to the abolitionists, from nineteenth-century reformers to contemporary political activists, fears about the mixing of alien ideas and the incorporation of foreign bodies have posed risks that are often likened to the digestive processes associated with food and eating. Chad Lavin (2013) goes so far as to say that America's obsession with diet and obesity can be understood as a response to perceived threats to individual and national sovereignty, while Probyn (2000) and Korsmeyer (2011) both reflect on the wider political and cultural significance of embodied notions of culinary desire and disgust.

What becomes clear from the wider literature concerning the moralization of food is that particular constituencies were regarded as most in need of moral guidance. Among these were the poor, especially those with families, which facilitated the surveillance of children at various sites, including clinics and schools and in the home, where women became the focus of the domestic science movement in Australia, the UK and US (Biltekoff 2002; Coveney 2000).[4] Interested in the cultural politics of dietary health, Charlotte Biltekoff (2002) has suggested that the movement not only provided a framework through which citizens might construct their identities as moral individuals, but in doing so also 'naturalised class differences and normalised a middle class standard for alimentary subjectivity' (ibid., p. 61). At a time when science and technology had acquired an aura of divinity (Shapiro 2009), domestic scientists 'fashioned a body of empirical and moral truths through which modern subjects could come to know themselves and their place in a social order increasingly characterised by proper eating' (Biltekoff 2002, p. 61). Consequently, not only did the domestic science movement perceive itself as providing a body of knowledge that could facilitate self-making but—importantly—it also enabled individuals to evaluate and classify others (ibid., p. 64).

Among those 'alimentary deviants' (Biltekoff 2002, p. 67) identified by the movement was a growing immigrant population bolstering industrial cities in the US at the turn of the twentieth century. In her examination of the use of ready-to-eat foods in cities such as New York, Chicago and Pittsburgh, historian Kathleen Turner (2006) documents the popularity of shop-bought bread and baked goods among European immigrants in the period between 1880 and 1930.[5] Eschewing the attempts of middle-class reformers to extol the virtues of a wholesome New England diet, working women took advantage of the pie-stands and bake-shops which sprang up in many industrial cities. While the reasons for this practice are numerous—including spatial constraints for baking and storing bulk-bought ingredients in small tenement apartments, the relative cost of buying the ingredients and fuel required for home-baking, as well as the cost of women's time which might otherwise be spent earning wages to support the household—Turner contrasts perceptions of the virtuousness of the housewife heroically producing 'honest' nutritious food from the most frugal raw ingredients, with the social stigma borne by their working-class counterparts, reported as 'poor and unintelligent', 'slovenly', 'lazy no goods'. These observations indicate that—insofar as 'food choices' are concerned—the ground was already fertile for the discourse of class pathologization to grow (Hollows and Jones 2010) as witnessed in more recent times—for example through Jamie Oliver's work in both the UK and US; a discourse that is disproportionately directed at women, who not only shoulder the burden of 'feeding the family' (DeVault 1991), but are also the perceived dietary custodians for the family.[6]

The latter half of the twentieth century witnessed an expansion in the moralization of food choice in the form of the 'alternative food movement'. Originating in the counterculture of the 1960s, the alternative food movement was based on a critique of the modern industrialised food production system (Belasco 2007). Involving a complex set of discourses concerning equality, 'naturalness', risk and anti-capitalism, this was a highly moral movement which sought to question and revolutionize the dominant social order by giving preference to local food over international trade, small- over large-scale production, organic over scientific farming methods, and to prioritize a vegetarian diet. As Belasco observes, the once 'alternative' ideals of this movement have perhaps not succeeded in radically changing the contemporary food system, but have nonetheless filtered through to public consciousness and, to some extent, have been reflected in changes in mainstream food production and consumption. This mainstreaming of the alternative appears to have occurred as a result of the food industry identifying business opportunities in the powerful discourses regarding 'sustainable development' and 'sustainable consumption', enrolling various critiques of technical and scientific innovations in food production (such as pesticides, GMO, and 'plasticulture') as well as debates concerning the fairness of wealthy consumers in the Global North thriving at the expense of those living elsewhere, including distant farmers in the Global South. Indeed, eco-labelling and certifications of production methods,[7] as well as more diverse spaces of exchange, including the proliferation of alternative food networks and farmers' markets, have contributed to an increase in self-consciously ethical consumption among those constituencies who can afford to make such choices.[8]

Having established how food has come to occupy a moral terrain in advanced industrial nations, we now turn our attention specifically to convenience foods, a vehicle through which a number of moral conflicts and ambivalences emerge. It is often suggested that while the state has declared a commitment to promoting healthy eating across numerous nations, information about food and health has nonetheless been shaped by the commercial considerations of the food industry. This has often resulted in the individualization of responsibility and the attribution of blame to those who might otherwise be considered as victims (cf. Keane 1997; Askegaard et al. 2014; Evans 2011; Meah 2014).

De-moralizing Convenience Food: From Convenience to Care

In a critical reflection on 'care' and its application in food studies, Anna Lavis et al. (2015, p. 20) describe the term as diversely applied but generally associated with values such as 'commensality, affection, love, kinship and social cohesion'. 'Care' is, according to the authors, used to denote feelings of concern for food and what people eat, as well as various forms of work where food is linked to sociality, to the manifestations of belonging and the formation of social groups. By linking care through food to the performance of self and group identities, Lavis et al. elucidate how notions of 'proper' or 'improper' ways to care can be critical to understanding the moralization of convenience food.

Mirroring the language used to describe those who used earlier forms of convenience foods (as reported by Turner 2006), Gofton (1995) examines the etymology of 'convenience', suggesting that the word has traditionally been associated with laziness, immorality and unhealthiness. Not only did this offend middle-class sensibilities framed by Christianity, on the one hand, and the domestic science movement, on the other, but—additionally—Gofton suggests, it presented a potential threat to the social order (ibid., p. 159).

Nonetheless, convenience foods have a key place in culinary history; indeed, they played an important role both during and after the Second World War, a period characterized by food shortages and a reliance on culinary innovations such as powdered eggs and milk, along with tinned meat. It is perhaps no surprise that, in the 1950s, the American food industry was anxious to expand the use of pre-prepared, packaged ingredients that would appeal to working women who had been 'liberated' from the home, women who—it was claimed—'loathed' cooking (Shapiro 2004). However, as culinary historian Laura Shapiro (2004) notes, there was a lack of independent research to corroborate this claim.[9] Instead, a series of projective tests carried out by behavioural scientist, Mason Haire, revealed how the use of convenience foods was viewed at that time. In these tests women were invited to express their views about fictional shoppers, whose shopping lists initially included freeze-dried coffee and, in a later test, blueberry pie mix. Their responses produced 'devastating' results: instant coffee users were described as lazy, disorganized, spendthrift and as bad wives (ibid., p. 54), while those who used the pie-mix fared no better. Haire's main interpretation of these results was that the use of prepared foods was not consistent with the role of a good housewife (Lane and Watson 1975, p. 49). Clearly, there was a need to disentangle 'short-cuts' or 'convenience' products from their slatternly image, which, Shapiro writes, required a radical reconceptualization of the meaning of the verb 'to cook' (2004, p. 55). Aided by advances in technology and a burgeoning availability of food items, food marketing aimed at middle class women attempted to reconstitute cooking as 'fun', and all kinds of foods—including the 'TV dinner'—were re-inscribed as integral to a post-war image of modernity and progress (Inness 2001, p. 157, see also Warde 1997). Moreover, these goods were depicted as expanding women's culinary repertoires, rather than narrowing them (Inness 2001, p. 159, see also Shapiro 2004). Indeed, Belasco (2008, p. 55) goes so far as to argue that the whole industrial food system markets one simple product: convenience.

Given the moral disapprobation with which convenience food has historically been tinged (Warde 1999, p. 518), it is unsurprising that food advertisers would want to distance such goods from the pejorative connotations of convenience, frequently positioned as the opposite of 'homemade' with its positive associations with love and care (cf. Warde 1997). Preferred synonyms identified in popular magazines from the 1960s onwards included speed and ease of preparation. Rather than seeing convenience as the opposite of care, however, advertisers sought to reframe convenience foods as vehicles through which women might successfully mobilize a discourse of love or care. Their products have 'love value' (Parkin 2006, p. 30): the power to capture hearts, express love, hold marriages together and secure the approval of capricious children (cf. Warde 1997; Burridge and Barker 2009). Importantly, food advertisers were careful not to suggest that their products surpassed the qualities of homemade food. What they did do, however, was to acknowledge that while homemade is best, there are times when the judicious use of particular pre-prepared food items might enable one to successfully perform one's culinary duties.

In a discussion of the culinary antinomies of care and convenience, Alan Warde (1997, p. 132) offers an interpretation of a 1991 advertisement for Stork frozen pastry products that appeared in *Woman's Weekly* via which, he suggests, one is invited 'to feel less uneasy about not expressing love properly'. While it is make-believe, counterfeit and inauthentic baking, it nonetheless fulfils some requirements—expressing love and smelling of fresh, real baking—and is, therefore, acceptable since there will be other occasions when one will make one's own pastry. So, too, Carrigan and Szimigin (2006) argue, were women able to exert their power as 'mothers of invention', combining 'fresh' and 'convenience' foods in artful ways to preserve their self-image as caring parents.

Such discourses are no less apparent today, with the notable difference that women are no longer imagined as the sole audience. Indeed, English-based food company, Colman's, has extolled the virtues of its dried packet sauces as having the capacity to help male and female cooks make 'proper British meals that say it all'. This is illustrated, for example, in a TV advert where a father suggests that 'this shepherd pie says…' a myriad of comforting things that he is unable to articulate to his heartbroken, newly-single teenage daughter. The British supermarket chain, Morrison's, also launched a television advertising campaign in which it claims: 'Love. Morrison's makes it'. In one of its advertisements, broadcast in 2016, a ready meal with pre-prepared broccoli and sliced carrots (of unknown provenance) featured centrally in the performance of love by a father to his partner returning home from a night shift.[10]

These advertisements offer public articulation of private domestic practices which involve the combination of processed, pre-packaged or pre-prepared items with a range of 'fresh' ingredients, consequently blurring the definition of 'homemade' which increasingly emerges as a socially constructed and—perhaps—outdated ideal. Indeed, even among Swedish mothers engaged in weaning, the most powerful arbiter of their achievement of a successful mothering identity is no longer whether or not they prepare homemade food for their babies (Brembeck and Fuentes 2017). Instead, in this context, food choice is rationalized by obligations to have food of appropriate quality and provenance 'at hand'. Serving commercial baby food that was 'tailored' for babies' special needs was described as just as appropriate and responsible a way of caring as offering homemade baby food. Indeed, eco-labelled and locally produced commercial baby food was described as a way to care for the baby while at the same time caring about the environment.

Examples such as these illustrate why it is important to avoid over-drawing the distinction between convenience foods and those which are homemade, 'properly with love, from scratch' (Slocum et al. 2011). Indeed, Marshall and Bell (2003, p. 62) have insisted that rather than seeing convenience and home-made foods as separate categories, they should instead be understood as part of a continuum, distinguished by context not content. Similarly, Warde (1999, p. 519) suggests that convenience food is 'not just a set of properties of food items but … a matter of social context'. Following calls for more ethnographically-informed research on consumer practice (Jackson 2016), the remainder of this chapter focuses on the social contexts in which decision making about food is made.

Although empirical evidence already exists concerning the ways in which convenience foods are used, appropriated and made sense of in everyday life (as reviewed by Jackson and Viehoff 2016), much of this work has focused on discursive recall and reported practice. Here, our understandings are informed by observations of the 'doings' as well as the 'sayings' of social practice (Schatzki 2002) which help contextualize and make sense of what people say and do around food and—importantly—why. Our discussion draws upon data concerning the use of convenience foods in everyday cooking and eating in England and Germany, and the use of commercial baby food in Sweden. While many participants indicated an awareness of how convenience foods have become morally framed, our data were also characterized by justifications which, rather than suggesting that convenience and care are conceptual opposites, pointed toward the enrolment of convenience foods in the broader dynamics of caring. Indeed, for many of our participants from across a range of social and economic backgrounds, convenience food was articulated as a vehicle through which care could be expressed; for example, by enabling them to accommodate the needs and preferences of different family members, to demonstrate thrift and economy in managing financial resources, or to help avoid food waste and consume more sustainably. Our empirical work with consumers is full of moral justifications, 'unapologetic apologies' (Meah and Jackson 2013), irony and self-deprecating humour, underlining the attribution of ethical significance to the consumption of convenience food, while also expressing moral ambiguity regarding its use. We begin by exploring how convenience is reconstituted as a means through which to express care.[11]

Reframing Care via Convenience Foods

That everyday provisioning is a complex practical and emotional endeavour which has to fit the circumstances of people's everyday lives is well documented (cf. Carrigan et al. 2006; Bava et al. 2008; Meah and Watson 2013). Those responsible for provisioning—shopping for and planning meals, preparing and disposing of food—must fit these responsibilities into complex repertoires of domestic life which often involve balancing paid work outside the home, childcare duties—including transportation to and from childcare—as well as children's educational and leisure activities, external caring responsibilities, and social or leisure commitments, along with having to consider the tastes and preferences of a number of family members who may or may not be permanently resident (for example, grandchildren, people who work or study elsewhere and children who are parented across multiple households), all of which must be achieved within a given financial budget. All of these were concerns that manifested themselves in how participants—across different countries and in differing financial circumstances—contextualised their practices. Drawing upon Simmel's ideas about value—which is not articulated as an inherent property of *objects*, but as a judgement made about them by *subjects* (cited in Appadurai 1986)—we examine how individuals acknowledge and evaluate moral discourses and transform them by engaging reasoned and practical logics in making sense of and qualifying their use of convenience foods (cf. Sassatelli 2004).

Being Sustainable, Reducing Waste

Talking with our participants, it became apparent that different scales or geographies of responsibility (Massey 2004) were invoked in discussing their reasons for using particular convenience foods; in differing ways, these could be interpreted as morally motivated choices. For some, buying pre-prepared ingredients or meals was described as a way to consume in a safe and fairer way. Klara, a 31-year-old first time Swedish mother, described eco-labelled commercial baby food as a convenient way to consume in a manner that allowed her to care for her baby and for the environment. Expressing great concern for the environment and animal welfare, as well as the safety of the food she gives to her son, Klara reported how she always opted for eco-labelled and vegetarian products when purchasing commercial baby food. For Klara, and others like her, buying baby food produced in Sweden reduced 'food miles' and helped address concerns they might have regarding the safety and quality of ingredients and modes of production. However, many of the women's stories parallel Meah and Watson's (2013) findings regarding the ways in which consumers' concerns about provenance are embedded in everyday practices of provisioning, characterized by complexity, ambivalence and trade-offs. Although provenance was imagined as a way to care for a range of distant others (farmers, animals and future generations), the decision to buy eco-labelled baby food, as reported to Maria Fuentes, could also be related to the baby's preferences:

Klara: We have made an effort to buy eco-food. Hipp has a lot of them. Well, I don't know, I think he would prefer them anyway. For example, Hipp has a veggie lasagne and he finds that one much tastier than Semper's that isn't eco-labelled. We did try that one too, but well … the favourite jars are corn and sweet potato, the eco-ones.

Maria: So he has favourites?

Klara: Yes, but of course. When we started [feeding him] he really liked those. And then we thought 'OK, we give him the ones he likes to help him to get used to textures and so on'.

Among other participants, buying pre-prepared ingredients enabled them to avoid wasting food. For example, while his wife implies an assumed moral judgment in the statement, 'I know it's not the best way to buy things', Edward, one of our British participants, was unambiguous in justifying his household's use of bagged, washed and trimmed root vegetables for use in casseroles and stews which he articulated as a means of demonstrating care for others who are hungry:

> It has everything you want in, so there's no waste … it's got everything you need … I don't like to waste things … It's disgusting throwing food away when some people are hungry. No, I can't stand it.

A more common justification—discussed in Daniel Miller's (1998) account of making love in supermarkets—was a concern with the more immediate scale of the household, particularly in relation to financial resources. This was apparent across households with a range of incomes. Laura, a retired professional, for example, bought a combination of fresh blueberries and frozen mixed berries to eat with her breakfast cereal. Although she and her husband enjoy a comfortable income from their combined pensions, in stating that it seems 'extravagant' only to buy fresh berries (which might perish before she can use them), Laura implies that buying a frozen alternative is a better choice, at least at the level of household economy. At the poorer end of the economic spectrum, Tameka (a 28-year-old Black British woman of mixed African and Caribbean descent) also described how she combined fresh, frozen and canned ingredients in her cooking. Frozen chopped onions, for example, were experienced as a boon within her domestic economy since these (along with other frozen vegetables) gave her confidence in being able to have vegetables in the house, knowing that they will not perish in the way that fresh items might if unused before her next monthly online shop. She reports: 'I'm in love with using frozen veg as I find I don't waste vegetables and they don't go off … it really hurts me when I waste vegetables'. As a single, working mother, Tameka can ill-afford to waste any of the household's already limited financial resources. The role that financial resources play in the choice of convenience food is worth exploring in more detail.

Keeping a Budget

Our findings echo those of Henderson et al. (2009) whose work with South Australian consumers draws attention to the fact that 'bad' dietary decisions should not be assumed to emerge from poor health literacy. The authors contribute to an increasingly vocal critique of the 'deficit' understanding of consumer behaviour which isolates individuals and the food they consume from the social, cultural, structural and practical complexities in which everyday practices are embedded (cf. Halkier and Jensen 2011). Henderson et al. go on to problematize the moralization of 'food choice' by highlighting how some individuals consciously reject the primacy of health as a means of rationalizing decisions about what they 'should' be eating. Indeed, for people on low incomes, cost is reported to be a key determinant of purchasing decisions. This was certainly borne out in our work with some people who were living in precarious domestic situations and on very low incomes. Among them was Rageh, a 30-year-old Somalian man who has been seeking sanctuary in the UK since 2007. With his refugee status unproven, he was unable to claim state benefits and was, instead, living in temporary, shared charity accommodation, receiving just £10 (approx. €11) each week to live off. Accompanying him shopping and observing him cook revealed that Rageh was only able to afford one meal a day. This consisted of a monotonous cycle of cheap, frozen, low-quality chicken, combined with fresh onions and some frozen mixed vegetables, seasoned with spices and generally served with rice, pasta or maize. Aware of what the current nutritional guidance is regarding a balanced diet, Rageh explained:

> The thought of buying fresh carrots, fresh peas and fresh cauliflower never crossed my mind. It's so expensive and I don't think they would last as long as my frozen. It's all about quantity and cost, not quality or nutrition.

For Rageh, a concern with ensuring that he has a sufficient quantity of vegetables, acquired within his budget, overrides any concerns he might have about their nutritional quality. Rageh illustrated the cost/quality equation by explaining how different kinds of processed potatoes differ in cost from their raw form:

> Instant mash is cheaper than potatoes, even frozen chips are cheaper than potatoes. I can get three packs of instant mash for the price of one bag of potatoes.

He adds, 'instant mash will also save time in the kitchen', thus highlighting how a concern with saving time and effort is not necessarily about 'laziness', but an important consideration if one has limited physical energy caused by a lack of food.

In another example, Phil (a 44-year-old recovering alcoholic), unambiguously confirmed that financial considerations are his biggest priority when shopping. Drawing upon moralistic language, Phil admits that his 'downfall' is chocolate and cheesecake. In order to afford these luxuries on his income from state benefits, he reports that he will shop at the local market, where he can buy 'seconds' of a branded chocolate cake bar. Where Phil is potentially more exposed to moral disapprobation than most people who would admit to a weakness for chocolate—including the researcher (Angela Meah) who interviewed him—is that he has Type-2 diabetes. He is not, however, ignorant of the implications of his dietary decisions and poses a valid moral question: 'Where's the harm in that? OK, it's harming *me*.' Phil implies that private decisions about what to eat are a concern for the individual alone, not least when contextualized in terms of his previous addictive practices concerning alcohol which affected both him and his loved ones.

Accommodating Dietary Tastes and Preferences

It is perhaps not surprising that one of the biggest challenges in 'making love' through one's provisioning practices (Miller 1998) can be observed in attempts to accommodate the tastes and preferences of different family members. In the discussion that follows, however, rather than focusing on convenience foods as a cost-effective means with which to feed one's family, we focus instead on how different kinds of food have different types of love-value for our participants; value that might not be appreciated consistently across all family members. Sometimes love might be directed toward the self, particularly in those households with older children whose preferences were prioritised when 'crap' items (typically confectionary and savoury snacks) found their way into the shopping trolley. In our research, men were likely to have more 'adventurous' palates than their partners or children, and numerous examples such as pickled salsify, canned soup, curried ready-meals, tins of cassoulet, confit de canard or pâté were presented as indulgences which 'nobody else would want to share particularly, or approve of either' (as one participant remarked).

Given the visibility of children in discourses of morality (cf. Coveney 2000), babies and younger children emerge as particularly important figures in negotiating 'food choice' within families. However, as was pointed out to Jamie Oliver by some of the Rawmarsh mothers, some children will only eat what they like, and what children want to eat may not conform to top-down food policies formulated to tackle the problem of obesity (cf. Fox and Smith 2011).[12] A further consideration is how parents' moral responsibility to ensure that their children are exposed to positive 'food choices' is also situated in a broader discourse which positions the child as citizen, a framework for which was laid down in the UN Convention for the Rights of the Child (1989). As Coveney (2000) states, the new ethic of parenting requires that children should be given choices *and* that they eat nutritious foods. This has resulted in the emergence of trends such as baby-led weaning, which is particularly popular in the UK (Keenan and Stapleton 2014). Among producers of what may once have been characterized as 'morally inferior' commercial baby food, the importance of the child as a competent actor in the weaning process is reflected in both product development and marketing strategies. As Fuentes and Brembeck (2017) observe, some baby-food producers have designed packaging to be accessible for small hands to manage, while product marketing claims that 'the love of good food is developed by meals that are both scrumptious and fun' (see, for example, https://www.ellaskitchen.se/, accessed 28 November 2017). Here, the authors argue, companies such as Ella's Kitchen (in the UK) and Semper (in Sweden) are invoking notions of 'fun morality', a phenomenon premised upon beliefs concerning a post-war morality via which children should enjoy life and follow their instincts, and mothers should enjoy motherhood and do fun things with their children (Wolfenstein 1951). The food products are designed and marketed as 'moralized products', that is, they are scripted or given qualities that afford certain actions of their users (Jelsma 2006).

Since powerful moral discourses surround motherhood, women in particular are exposed to the paradox of convenience and care, consequently rendering them more likely to construct themselves discursively in relation to discourses of good/bad or responsible/irresponsible parenting.[13] Elsewhere in the literature, however, is an emerging counter-discourse which reframes the use of convenience foods—from ingredients to take-away meals—as a means through which care can be expressed, whether this is in freeing up time to spend with one's family, or in providing an activity around which family-time coheres.[14] Discussing mothers' use of convenience products, Carrigan and Szmigin (2006) report that while some women's accounts continue to be characterized by guilt, for others, the inclusion of such items can be interpreted as no less eloquent a statement of love and care for one's family as the morally valorised preparation of meals from scratch. Indeed, they suggest that the use of convenience foods may be a strategy for mothers to enhance rather than detract from their devotion to their families (ibid., p. 1127).

A particularly poignant example of this can be seen from our work in Germany with Lena, a 37-year-old single mother of two children, aged three and a half and seventeen months. Knowing something of a participant's personal history can prove invaluable in helping to understand and give context to their practices. Lena reported that she is classified by others—and indeed sometimes refers to herself—as 'asi' (*azsocial*/anti-social) which carries the same kind of negative associations as the word 'chav' in the UK (Tyler 2008). Lena reports a history of domestic abuse during her upbringing and adult life, with part of her childhood having been spent in social care. Perhaps because she lacked positive maternal role models in her own childhood and because the children's father was absent, Lena has weekly support from a social worker. However, she rejected the suggestion that she needs parenting advice: 'nobody needs to tell me anything'. Not unlike Phil (discussed above), Lena is aware of what a nutritious diet looks like, but this has to be negotiated within the limitations of her budget and what her children will eat. Speaking of her older child, Birk, Lena explained: 'you often have to force him, you know, that's why I always buy what he really eats'. Sometimes this has brought her into conflict, with adjurations from Birk's nursery regarding acceptable sandwich fillings in children's lunches, but even the nursery had conceded that a more pragmatic response was required since: 'you can be glad he's eating anything'.

As the researcher (Valerie Viehoff) accompanied Lena and her youngest child shopping, it became apparent that Lena's provisioning practices were underpinned by her desire to be a 'good' mother, whether this manifested itself in paying for what she deemed to be overpriced 'children's apples', being led by her children's tastes and preferences, or setting rules around mealtimes to ensure that her son eats. Being a 'good' mother was also facilitated through the use of ready-to-bake bread rolls. This required Lena to get up early in order to prepare the rolls, something she found particularly stressful if she had had a 'rubbish night' [having slept badly]. However, her son's joy in knowing that she had done this for him outweighed the stress and inconvenience of an earlier-than-normal start to the day. The ready-to-use dough might not carry the same kind of love-value that homemade bread has, but for a harried single mother with two young children, it was a means through which love and devotion could be expressed and which is reciprocated in the child's anticipated response: 'I know that my son will rejoice and be happy'. We do not know whether Lena's son is happy because he likes the taste of the freshly-baked bread rolls or because he knows that his mother has cooked them for him. Either way, they seem to enhance the reciprocal relationship between mother and child.

It is also worth noting that it was not only mothers who deployed food as a vehicle through which to achieve 'successful' parenting identities. Imran is a 45-year-old Pakistani man whose work as a health scientist positioned him as an 'expert' on the relationship between nutrition and health. However, his wife, Maryam, complained about his shopping trips with their three children—aged nine, twelve and fifteen—from which they would return with confectionary, cakes, biscuits and crisps, items which, she reports, she would never buy. Maryam speculates that Imran buys these things for the children 'to make them happy' and to enable him to indicate—to himself at least—'I'm a good father'.

Like Lena, the couple's eldest child is very particular about what he will eat, preferring take-away food to his mother's cooking. She reports that her husband is content for him to eat confectionary and crisps since these will at least 'give him some energy'. Each parent clearly has different conceptualizations of how to care for their son and Maryam reports that, for her husband, the priority is to ensure that—even if it is a bar of chocolate and a glass of milk—'at least he's eating something' before he goes to school. In this context, convenience products are seemingly valued—by Imran at least—as a necessary means for him to perform his responsibilities as a caring and concerned parent.

These data, from households in different European contexts and in differing circumstances, point toward a shift in how convenience food is conceptualized. Not regarded as a 'morally bankrupt' form of cooking associated with sloppiness or a lack of care for the health and well-being of one's loved ones, its significance can, in some circumstances, be reframed as a form of care. This might include freeing up time to spend with one's family, being more economical and less wasteful than buying 'fresh' or unprepared ingredients, or keeping different family members happy by catering to their diverse culinary preferences. Here we are reminded of Bugge and Almås's (2006, p. 220) suggestion that 'dinner cannot be said to be good or bad in itself, but only in relation to the values it is meant to realise'.

Negotiating Moralization with Humour

Given how morally loaded buying, preparing, eating and disposing of food has become and how the expansion of food availability has increased opportunities to do the 'right' or 'wrong' thing—as defined by various expert discourses—one might expect our participants' narratives to be loaded with self-judgement or expectations of judgement from others, including the researchers involved in these studies. While there was evidence of assumed judgement by imagined audiences, this was—in the UK in particular—often demonstrated through participants' use of irony and self-deprecating humour. In examining how the deployment of humour helps us to understand the ways in which individuals internalize or respond to the potential moralization of convenience food use, we draw upon the theories of humour reflected in contemporary academic scholarship (cf. Watson 2015).[15]

Acknowledging the context-bound and subjective nature of humour renders its interpretation a complex endeavour (Atkinson 1993; Bremmer and Roodenburg 1997). The relationship between humour and laughter is also acknowledged to be contested as laughter can occur without humour and not all humorous situations generate laughter (Shaw 2010). Here we explore how several categories of humour, including superiority, incongruity and relief, feature in how some of our participants talked about convenience food.

Superiority

The superiority theory suggests that humour is found in the misfortune of others, sometimes referred to as *schadenfreude* (Watson 2015). Applied to research into food practices, this might, for example, manifest itself through the mockery of other people's lack of taste, taken as evidence of the superiority of 'us' over 'them' (Janhonen 2017). Examples can be seen in studies of audience reactions to the Rawmarsh mothers featured in Jamie Oliver's TV programmes (Hollows and Jones 2010; Fox and Smith 2011; Piper 2013) and in the example (cited in Chap. 5) or Nicola's awkward laughter and expression of disgust over her mother-in-law's taste for processed mashed potatoes.

Incongruity

In another household, there was evidence of a discrepancy between reported and observed behaviour. The humour that arises from such awkward juxtapositions can be interpreted via incongruity theory. While it is not unusual to find a gap between people's doings and sayings, the example below was particularly amusing for all concerned. Ted, a retired professional, has taken part in several of our previous research projects. When the researcher (Angela Meah) explained the focus of the present study, he asserted that the term 'convenience food' was not something that really applied to their household—although when she gave examples from the full spectrum of goods which potentially come into this category, including items such as frozen puff-pastry which she had previously observed him using, he conceded that they do use particular items such as this: 'occasionally, we don't do very much with pastry'. While preparing a dish made with chicken breasts, Ted pointed to the chicken and observed:

'That's fairly convenient. That's the closest we get to convenience foods, chicken already breasted and boned', simultaneously pointing out that his wife buys these while he would buy a whole chicken. A discussion ensued during which Ted justified the use of various items, often invoking the approval of some of his favourite Mediterranean chefs in doing so. Ted's claim about not really using convenience foods unravelled a little later in the research when, on looking inside the refrigerator, he found that he had run out of preserved lemons (which he makes himself). Seeing the irony of the situation, he walked from the fridge to the larder, admitting: 'Now you've caught me in a very embarrassing moment … having bought a jar of preserved lemons [laughs]' which he uses as 'back-up' for occasions such as this. In this example, Ted does not distance himself from convenience foods on the grounds that they imply a 'lack' of time, knowledge, skill or effort—nor that the shop-bought lemons are inferior in quality or taste. The humour arises from the incongruity between what he says he does and what he is observed to do by the researcher to whom he has made such public protestations—a contradiction which he laughingly acknowledges.

Relief

Elaborated by Freud, the relief theory of humour applies to situations in which laughter is used to release some kind of awkwardness or tension, be that emotional or psychic. Evidence of this is widespread in our research. For example, it was noticeable in one UK household where one of the participants, Gloria, a white working-class woman in her forties, appeared to be particularly self-conscious about her food provisioning practices. Unlike some of our other working-class participants, Gloria seemed particularly nervous during the initial interview, when she appeared to be anxious about being judged. For example, the following extract reflects on the previous evening's meal:

Gloria: Yesterday I made chicken pie, that was fresh chicken and I put a stock cube in it and I put instant gravy, and I add a packet of sage and onion stuffing to it, but we had fresh leeks in it and then I bought some puff pastry and put that on the top. And then we had [laughs] it sounds as if we eat convenience food all the time…

Angela: *The way you said that then, you make it sound like you're expecting to be judged?*

Gloria: Yeh, yeh.

Talking about a previous meal prompted the recognition, from Gloria, that much of the food involved pre-prepared ingredients; her laughter defuses any embarrassment on her part, prompting reassurance from the researcher. On another occasion, Gloria begins filming by asserting 'today, we're having convenience food' and is very direct in explaining what the meal will consist of, as well as why: 'I've been to the gym; I'm tired [laughs]'. While her meal choice requires no justification or apology, Gloria nonetheless feels obliged to explain it, followed by laughter, which could be interpreted as nervousness or embarrassment. When she was subsequently asked who she imagined the audience for the video might be, she did not mention the researcher but 'other people who might be watching' and, presumably, judging.

These examples raise important questions about the role of the researcher in shaping the participants' responses. Participants, perhaps unconsciously, anticipate the extent to which the researcher shares their frame of reference and will 'get the joke', sometimes apparently fearing that their actions will be judged according to prevailing discourses of healthy eating or maternal care. The conversation between researcher and participant builds boundaries of inclusion and exclusion, highlighting the morally-charged nature of 'feeding the family', particularly where the use of convenience food can be seen as a potential abrogation of maternal responsibility or as evidence of a lack of care.

Conclusion

This chapter has explored the many ways in which contemporary food choices are moralized. What individuals eat has always been regarded as an indication of their moral character. But the moral implications of food choice have proliferated as those choices have become more abundant. Contemporary consumers are faced with a range of moral dilemmas about the nutritional qualities of what we eat, how our diet impacts on our health and well-being and on the environment. These issues are highlighted in the context of modern family life, where working-class mothers are particularly subject to the moralized discourse of 'feeding the family'.

The moralization of convenience food is manifested across a range of different scales (from concern about distant others to care for one's immediate family) and in relation to both human and non-human actors (in relation to Fair Trade and animal welfare, for example). In some sections of the media and in political rhetoric, convenience foods have become a synonym for all that is 'bad' and 'uncaring' about modern systems of food provision, involving intensive and unsustainable modes of production that are harmful to the environment and levels of consumption that are harmful to human health and wellbeing. Our research with consumers in a range of European contexts has problematized these assumptions, reframing the relationship between convenience and care, and exploring the many ways in which food is valued. Above all, our research seeks to avoid the moralization of 'choices' that are often deeply constrained, probing the context through which we might understand how these choices make sense to our participants, and searching for alternative frames of reference to those that assume a lack of knowledge or understanding. We take a similar approach in the next chapter which seeks to reframe the easy assumption that the increasing use of convenience food is evidence of a lack of cooking skills and competences.

Notes

1. Rawmarsh is a small town, with a predominantly White working-class population, located near Rotherham in South Yorkshire. With an industrial heritage concentrated on coal mining and steel production, the town has been characterized by high levels of unemployment, poverty and health inequalities following the decimation of those industries

toward the close of the twentieth century. Although located on different continents, the choice of Rawmarsh to some degree mirrors the selection of Huntington, West Virginia, as the site of *Jamie Oliver's Food Revolution* (Oliver et al. 2009): both are poor White towns, positioned—in the media—in relation to a White, middle-class, non-obese norm (cf. Slocum et al. 2011).

2. Jamie Oliver called one of the Rawmarsh women 'a big old scrubber', 'a fucking arsehole', 'a tosser' and 'an idiot' (quoted in Rich 2011, p. 16). In an infamous outburst on his TV show *Return to Jamie's Dinners*, Jamie Oliver declared that 'It's kind of time now to say, you know, if you're giving … very young kids bottles and bottles of fizzy drink you're a fucking arsehole, you're a tosser. If you give them bags of fucking shitty sweets at a very young age you're an idiot' (quoted in Rousseau 2012, p. 15).
3. Although religious and secular moralities have largely been supplanted by the new moralities of 'healthism', a study of health beliefs among older Irish adults (Delaney and McCarthy 2014) suggests that while Catholic practices of fasting and abstinence may have declined, religious moralities which can invoke feelings of guilt or shame persist in how some older people speak about food.
4. The development of baby clinics during the first half of the twentieth century implied a shift where mothers' knowledge was supplanted by 'expert' knowledge extolled by medicalized discourses of baby care and child development. While mothers were still responsible for the health and wellbeing of their children, they were nonetheless deemed in need of scientific and medical advice to fulfil that responsibility (Apple 1995; Coveney 2000).
5. Turner also examines delicatessen food, which—like baked goods—could simplify meal preparation at home, as well as ready-to-eat food available from pushcarts and food-stands or in saloons and small restaurants.
6. This responsibility is shaped by powerful, moralized bio-medical, scientific discourses directed at women during pregnancy and early motherhood. Mothers are consequently confronted with an array of socially and culturally dominant values concerning, for example, their own size before, during and after pregnancy (Fox et al. 2009; Keenan and Stapleton 2010), acceptable weight for infants and children appropriate to their age/developmental stage (Coveney 2000; Jain et al. 2001), as

well as prescriptive guidance regarding breastfeeding (Marshall et al. 2007) and weaning (Brown and Lee 2011; Moore et al. 2012).

7. Marketing devices such as packaging and labelling inscribe products with morals (Fuentes and Fuentes 2017). These 'market offerings', inscribed with notions of care for the environment, for farmers and for animals, illustrate how morality is not just in the minds of people, but also in things that afford the actions of consumers (Jelsma 2006).
8. In our earlier research on provenance, carried out with UK consumers (see Meah and Watson 2013), some participants made the point that 'morality is a privilege of the rich, to some extent'. Given the price premium often attached to ethically produced food items, one might argue that inaccessibility to all is itself a moral issue since 'choice is not actually there' for people on low incomes.
9. This is a recurrent theme in the moralization of food, as Murcott (1997) notes in the context of debates about the decline of 'family meals' where political rhetoric and media commentary have frequently run ahead of the evidential base.
10. For further discussion of these ads, see Jackson (2018).
11. A longer version of this argument has been published elsewhere (Meah and Jackson 2017).
12. For a broader discussion of how the moral policing of families has evolved in Australia and the UK, see Coveney (2000).
13. For an extensive discussion of the moralization of parenting practices, see Coveney (2000), Carrigan et al. (2006), Costa et al. (2007), Marshall et al. (2007), Tyler (2008), Fox et al. (2009), Halkier (2009) and Keenan & Stapleton (2010).
14. For example, de Boer et al. (2004) discuss Irish mothers' take-away food as a treat, while Meah and Jackson (2017) illustrates how pre-prepared ingredients can facilitate the participation of children in cooking, thereby providing a fun activity. Brembeck's (2005) study of how Swedish mothers incorporate fast food from McDonald's within culturally-approved ways of spending 'cosy time' with their families is also relevant here.
15. Browne (2016) has written about the role of humour in the context of debates about environmental sustainability which are often characterized by an earnest tone, marking the seriousness of the issues, while Janhonen (2017) discusses the role of humour among young people in relation to school food.

8

Cooking and Convenience

Introduction

The question of how and in what ways food provisioning can be made more convenient is closely linked with how cooking is understood and practised, since convenience food is so often contrasted with meals cooked from scratch. In previous chapters, the normalization and moralization of convenience food has been argued theoretically and illustrated empirically. In this chapter, we focus on the cooking practices associated with the use of convenience food.

These practices are often located within what might be called a 'discourse of decline' where there is a widespread assumption within political rhetoric and media debate that cooking skills have declined over the last generation or two. Initiatives such as Jamie Oliver's Ministry of Food were introduced in an attempt to reverse this apparent trend, providing simple recipes for people to follow or cooking courses for them to attend. Evaluating one such programme, researchers argued that 'there has been a decline in cooking skills [in the UK] since the 1950s'. The report continued: 'Home-made meals have been replaced with ready meals and convenience foods', concluding that 'This lack of cooking ability coupled

P. Jackson et al., *Reframing Convenience Food*,
https://doi.org/10.1007/978-3-319-78151-8_8

with ready access to high energy convenience foods has contributed to an increase in the number of people who are now overweight'.[1]

Despite these claims, the evidence of declining skills and their association with poor diet and long-term health issues is referred to by Lyon et al. (2003) as a 'modern myth', based on a handful of dated and inconclusive studies (cf. Lang and Caraher 2001; Short 2006). This chapter interrogates the relationship between cooking and convenience, seeking to challenge common assumptions and reframe current thinking. It focuses primarily on a case study of two Danish meal-box schemes as a particular kind of convenient food provisioning, making meal planning and shopping easier while retaining an emphasis on home cooking using the ingredients, recipes and cooking instructions provided. Comparisons are drawn with our other three cases (baby food, ready meals and canteen food) to raise questions about the skills and competences involved in this particular form of conveniencization. We begin, though, by clarifying what is meant by cooking.

What *Is* Cooking?

'cook' (verb): Prepare (food, a dish, or a meal) by mixing, combining, and heating the ingredients.
'cook' (noun): A person who prepares and cooks food, especially as a job or in a specified way. (Oxford English Dictionaries)

While dictionary definitions suggest that 'to cook' involves some kind of transformation of ingredients by the application of heat (cf. Lehrer 1969), academic research indicates that cooking is a rather more slippery concept. Indeed, Caraher et al. (1999, p. 595) have suggested that there is plasticity in what is meant by cooking. They question whether it must involve only raw ingredients, whether the key task is an assembly process and whether energy is required.

Since domestic cooking is something that occurs within private households, one might question what difference it makes—and to whom—whether cooking involves raw or processed ingredients, assembly, heat or any knowledge about how particular ingredients might work together. Coveney et al. (2012, p. 620) argue that the nature of cooking tends to become most visible when the skills implicated in its undertaking are thought to be in decline, found wanting, limited or inadequate, sometimes emerging as a 'moral panic'. In such circumstances, cooking has become problematized by 'experts' who privilege cooking from scratch (that is, cooking from basic or raw ingredients) as the benchmark against which all cooking should be assessed. Of course, debates about knowledge of cooking and domestic economy (particularly among the labouring classes) is not new (cf. Holm 2003; Lang and Caraher 2001; Coveney et al. 2012) and can be mapped alongside the emergence of nutritional science and the domestic science movement at the turn of the twentieth century (cf. Lang and Caraher 2001; Coveney 2000; Biltekoff 2002; Shapiro 2009; Coveney et al. 2012).[2]

Concerns about poor nutrition and poor household management were, however, displaced by matters of food insecurity during and in the aftermath of the Second World War, a period characterised by thrift and frugality (Coveney et al. 2012). Often eulogised as a 'golden era' of cooking, celebrity chef Jamie Oliver invoked the values of this period during the 2008 recession in his TV series and political campaign, *Jamie's Ministry of Food*. Although the original Ministry of Food was primarily concerned with providing information about how to cook with available rations, Jamie Oliver's TV series—also exported to Australia and the US—has been described as 'a grassroots response to the lack of basic food skills among adults', its central rallying call being to transfer skills among friends and family members by 'passing it on'. This intervention coincided with a period that witnessed the proliferation of TV cooking shows and associated lifestyle programming (see, for example, Hollows 2003a, b; Holden 2005; Feasey 2008; Swenson 2009), along with an increased appetite for 'nostalgia cookbooks' (Sutton 2001). The assumption behind Jamie Oliver's invocation to 'pass it on' was that the older generation was more skilful and accomplished in cooking than the younger one (Moisio et al. 2004) as implied by the 'discourse of decline'.

Based on his research on the Greek island of Kalymnos, David Sutton (2001) has described the acquisition of cooking skills as a form of 'embodied apprenticeship', learned through shadowing an older—usually female—relative, and then practiced, refined and improvised through repeated performance. If we acknowledge 'skill' as something which is developed through trained practice and embedded in routine, involving the conversion of information and practices into tacit knowledge (Sennett 2008), one might argue that a process of 'dis-embedding' has occurred in relation to cooking skill. While the alleged 'deskilling' of consumers has emerged as a dominant refrain within popular media and policy discourse in a range of national contexts, as with its partner complaint regarding the 'decline of the family meal', a number of scholars have cautioned against the easy assumption of a decline in contemporary cooking skills in the absence of convincing empirical evidence about the cooking practices of previous generations (e.g. Lyon et al. 2003; Murcott 1997; Short 2006). Indeed, Meah and Watson (2011) found that among their British participants, mothers were not uncommonly remembered as having limited cooking skills and a modest repertoire of regular dishes.

It has also been argued that domestic skills have been affected by the mechanisation of cooking, transforming the cook into an 'unskilled spectator who watches the machine function in her place' (Giard 1998, p. 212). This view has not gone unchallenged. Indeed, a number of scholars have highlighted that 'skill' involves more than simply understanding ingredients. Baking, for example, has been described as a 'science' since it relies on the precise measurement of specific ingredients.[3] Frances Short (2006) goes further in pointing out that baking a cake is more than just a mechanical action; it involves perceptual skills requiring judgement and understanding, not just of the ingredients but of the means by which they are transformed in the oven. Reporting on the introduction of the thermostat-controlled cooker and, later, the microwave oven—technologies assumed to have taken the stress out of cooking on solid fuel stoves or open fires—Elizabeth Silva (2000, p. 618) observes that recipe books produced for women users of these technologies did not assume an ignorant operator, but required agency and skilful assessment of the processes and results of using them. This is parallel to recent work on the skills required to cook with a multi-purpose cooker among contemporary Portuguese consumers (Truninger 2011).

Of course, changes in how food is produced, supplied and consumed from the 1960s onwards, including the proliferation of pre-processed foods ranging from those which are ready to use, ready to heat or ready to eat (Pfau and Saba 2009), has further complicated the question of what counts as 'cooking'. Indeed, while a range of concepts has evolved in popular usage—'home-made', 'home-cooked', 'from scratch', using 'basic ingredients' or 'ready prepared'—these remain largely unexplained, lacking clear and consistent definition (cf. Halkier 2009; Short 2006). While Short (2006, p. 51) draws attention to the lack of detail in academic accounts of the skills which are claimed as being lost and the processes by which these are said to be disappearing, the situation becomes even less clear if we consider recent empirical research which has attempted to explore what these terms mean to the individuals concerned.

Reporting on their work on 'perceptions of cooking' in Baltimore, for example, Wolfson et al. (2016) identify a continuum from cooking only from basic, unprocessed ingredients to any form of cooking in the home. For some, the latter might involve combining a packet of noodles with hot dogs and ketchup. Likewise, in their work with Irish consumers, Lavelle et al. (2016) report that for some, 'cooking from scratch' included the use of processed items (cf. Moisio et al. 2004). Indeed, Daniels and Glorieux (2015) have indicated the existence of a category of food products (for example, bread and dairy products) that are now normalised as staples but which might once have been considered convenience foods.

Short (2006) suggests that cooking is best understood more broadly as one of a myriad of household tasks (cf. Oakley 1985; Murcott 1995). As such, cooking should not be considered as a discrete activity, but one which has to be negotiated and organised within the possibilities and limitations afforded by everyday life. Reconceptualising cooking in this way is done through the practice-theoretical perspective which we adopt in this book.

Empirical evidence from a range of national contexts quite clearly points toward a need to expand what can be included in contemporary understandings of 'cooking' (Holm et al. 2016), emphasizing the complexity of the concept as well as related terms such as competence and skill. Lang and Caraher (2001, p. 10) argue that cooking is undergoing a process of rapid transformation and that the influences of globalization, changing production and processing methods, and the rapidly increasing availability of convenience and take-away foods must not be overlooked in determining what people do with food in the home. Indeed, they question whether the meaning of cooking should, in fact, be conceptualised as a process or an end-product (ibid., p. 12). The latter is popular among those concerned with public health and other forms of official guidance. However, in our research on cooking in everyday life, foregrounding the *process* highlights a rather more complex set of negotiations which are indicative of a wider set of skills and competences than implied in the straightforward transformation of ingredients with the use of heat. Not only would this require the inclusion of reheating a meal in the microwave or the preparation of a meal involving ingredients to which no heat has been applied such as salad or sushi, it would also involve an acknowledgement of a myriad of invisible skills, tacit assessments and often-unacknowledged sensory judgements which constitute the production of a meal. These include knowing how to rescue a sauce when it curdles on the stove, knowing how to make a dish go a little bit further, remembering that one family member will only eat a particular type of cabbage, or indeed knowing the app on your mobile phone that helps keep track of a shared shopping list in the digital cloud as some of our Danish participants did. Indeed, Ekstrom and d'Orange Furst (2001) suggest that the task of making dinner may begin before one sets off for work with a glance in the fridge or kitchen cupboards to determine what needs using up before it goes off and the additional ingredients one might need to go with this, or remembering to remove something from the freezer to defrost.

Shopping is increasingly acknowledged as an important dimension of food planning and preparation since it is at the point of purchase that consumers are required to trade-off a range of values including health, cost or convenience (cf. O'Brien et al. 2015; Meah and Watson 2013). It is also the point at which consumers begin to deploy a wider range of skills, including the ability to improvise. For example, Meah and Jackson (2013, p. 586) discuss how one of their British participants, while not being averse to using ready-made quiches when she could not be bothered to cook, knew how they could be improved by stabbing the top, filling it with an egg and embellishing it with additional toppings. The same participant explained how she would often bulk-out a shop-bought sauce by looking at and then adding to the ingredients listed on the jar. This type of improvised 'economy' was a skill learned from her mother.

These studies clearly problematize the suggestion that Western consumers have lost the knowledge of how to transform raw ingredients into meals. Instead, they indicate that knowledge about meal preparation has been reconfigured via a redeployment of knowledge and skill in other areas. These include planning and organisational skills, knowledge of when, how and where to shop and trading an awareness of health and nutritional discourses against more prosaic concerns such as what family members are prepared to eat, the range of other obligations which compete for one's time, as well as the financial resources available.

The Practice of Cooking

The empirical part of this chapter focuses on the cooking practices observed in the Danish case-study concerning the use of meal-box schemes to provide for evening meals in participants' everyday lives, although examples from the other empirical cases are drawn in as well. The Danish sample consists of families, with and without children, so our analysis may not apply to single-person households. The interviews, participant observation and focus groups were conducted by Frej Daniel Hertz and details of the participants, all of whom are anonymised via the use pseudonyms, are given in the Appendix.[4] The two different meal-box schemes to which the participants subscribed are respectively *Årstiderne* ('The Seasons'), which is an all-organic scheme, and *Ret nemt* ('Fairly Easy'), which includes conventional as well as organic foodstuff. Each box includes raw ingredients in precisely measured amounts, suitable for making specific meals for a set number of people, detailed recipes for each dish, with precise cooking instructions and cooking times.

Interpreting patterns of cooking from the meal-box schemes enables a comparison of how participants cook with these box schemes and how they cook without them, because none of the participating families subscribed to a meal-box scheme for the whole week, some of the families had entered, left and re-entered meal-box schemes over a period, and all the participants had years of experience with providing food and meals without using a meal-box scheme. Two introductory observations should be made in comparing cooking with and without the meal-boxes.

First, planning, shopping, storing, clearing up and disposal are intimately tied up with cooking, to a degree where it does not make sense to think of them as discrete practices, but rather linked activities within the compound practice of cooking. Apart from shopping, using recipes and clearing out the kitchen, the participants talk about and show how cooking is dependent upon a number of different food activities, such as using a mobile phone app to keep a shared shopping list, making one's own 'ready meals' and stocking them up in the freezer, using leftovers for next day's lunch or dinner, growing vegetables and herbs in the garden, having a division of labour between baking and other types of food preparation, and throwing out leftovers from an experimental meal which nobody in the family liked to eat.

Second, the interactions between cooking and other practices, such as parenting, employment obligations and the logistics surrounding leisure activities, are important contributors in shaping the conditions of how cooking can be done, whether or not it is achieved via the use of meal-boxes. Across other differences among the participants, it was practices with, or in relation to, children that were mentioned in particular and which could be seen from the observations. For example, how to cook while helping with homework, how to fit the logistics of feeding everybody while making sure that everyone got to where they needed to be for leisure activities, adapting evening meals to the unexpected influx of friends (or the equally unforeseeable failure of children to be home for dinner), making meal-times nice and cosy for children, and coordinating meal-times and eating with wider parenting practices. Work-life practices, including part-time or full-time employment, as well as the absence of clear boundaries between work and family life, also shaped how it was possible to cook. Finally, cooking overlapped specifically with cleaning, use of various media, and with activities such as baking, preserving and preparing cold packed lunches for family members which were distinct from main meal preparation

Cooking Without a Meal-Box Scheme

When looking at the ways in which participants cook when 'on their own', without the meal-boxes, three distinct ways of cooking can be identified. Each of these ways of cooking does not necessarily belong to one family or one individual. Rather, they are ideal types where, across different families, there is a tendency to shift between the three different ways of performing cooking, according to social interactions within the family, intersections with other kinds of practices, material arrangements and contingencies. The threefold typology can be summarised as:

- The planned way of cooking
- The improvised way of cooking
- The audit way of cooking

The *planned way of cooking* is the kind of cooking where participants or the main cook plan, usually for one week at a time, what the family should eat each evening, and which food items should be bought or otherwise provided for, in order to prepare these particular meals.[5] Here, Helene explains common meal planning within her household:

> During the weekend, we usually make a meal-plan for the coming week, so we find out who is at home which days, who is cooking when, and then we try to make a plan. And then simply write it into the calendar, Monday is chili con carne, Tuesday is tortillas, Wednesday is something else. And then we go out shopping on the Sunday, Paul and I, and maybe one of the kids.

The planned way of cooking does not necessarily actively include both parents:

> I probably tend to plan for myself what we are having, so all the time, I am in control of what is actually in the fridge, what has to be eaten first, and what is it that's easy to make with that. (Hanne)

As a part of the planned way of cooking, participants more or less stick to a structured meal plan:

> But perhaps we have become better at that thing about planning it. And that's because I can see that it works for us. You know, it's nice, so you don't have to use brain capacity on standing there and 'oh, what shall I come up with?' (Anja)

Thus, all ingredients are often there long before the actual cooking commences:

> And it does make everything much easier in everyday life that it's planned and shopped, and you know that you just have to go home and make it. (Betinna)

The cooking itself—turning ingredients into a meal—is timed carefully in the sense that there is an upper limit to how much time it takes. Practical procedures in cooking such as rinsing, chopping, heating, roasting etc. are linked explicitly with engagements in cooking such as making time for other practices, getting varied meals, respecting the tastes of other family members, making healthier meals and wasting less food. An example, illustrating the relationship between making time for other practices and cooking, comes from Jytte, who describes how she and her family chose a dish that would roast slowly while they went out for a run. We will return to arguments about ecology, health, and minimising food waste when describing how the participants cook *with* the meal-boxes in the next section. When cooking, the quantity of ingredients is adjusted according to the number of family members expected to be present. Either as a result of forward planning or inaccurate assessment, left-over food formed the basis of subsequent meals (such as lunch the following day): 'And there were left-overs, so each of us got a skewer for the lunch packet, instead of that boring sausage sandwich' (Anja). Alternatively, left-over items can be refrigerated or frozen to be reconstituted as another meal at a later date:

> A pie without a bottom where the main ingredients really are eggs and flour, and then you throw everything else from the fridge in, whether it's a squash or whatever it is. We have that fairly often. The kids are relatively happy with that. Sometimes a red pepper goes in, onion, squash, bacon or ham. (Rikke)

The *improvised way of cooking* refers to when decisions about what to cook occur on the day or the day before, be that as a negotiation between parents or a decision by one of them, depending on the division of food labour within the household. This way of cooking is, for example, enacted by Rikke when talking about preparations for cooking: 'I was standing there late in the afternoon—what are we going to eat?' Participants maintain a rough overview of what is already in the fridge, cupboards and freezer, and then shop for whatever is necessary to supplement the existing stock in order to produce a meal. One participant expresses it like this: 'Phew ... I never make meal-plans. I sort of shoot from the hip when I'm standing down there' (Gro). Another explained:

> Ehm, but ... then we run it a little bit ad hoc, you know. We had decided to make tortillas today, and then we hadn't got round to take the meat out of the freezer. Well, then I went past Netto on the way home and shopped for what we needed. (Helene)

As was also illustrated in some of our UK households, the improvised way of cooking enables cooks to utilise ingredients found in fridges, freezers and cupboards, some of which may be processed to various degrees, in addition to items which they might need to purchase to complete the dish. Where such decisions have to be made spontaneously, frozen, canned and dried products enable cooks to create a meal quickly or to eliminate unnecessary processes.

Most of the participants juxtapose weekend cooking with ordinary, everyday cooking. Here, Agnes explains her household's consumption of semi-processed goods: 'You know, in everyday life, if we are having salad there, then I've become a fan of the ready-made salads that you can buy'. However, this kind of supplementing part or complete meals within the improvised way of cooking is not necessarily reserved for meals consumed between Monday and Friday. For example, Marie suggests that 'sometimes at the weekend we can actually have some take-away'. Reporting an emergency situation, Maiken explains how the practical procedures involved in delivering a meal for one's family must ultimately be negotiated within and responsive to the exigencies of everyday life:

> And it's when you come home at 5 and have to whip up something ... you know this Wednesday, I got a flat tyre, and I had to phone Gregers and tell him to start cooking the food. If that had not been possible, to get him home and starting, we would have probably resorted to something else. You know, that does happen, that you are stuck somewhere. (Maiken)

The *audit way of cooking* is the kind of cooking where finding out what to cook is performed in dialogue with 'the target group' for dinner and it can be more or less planned. The key pattern is that the main cook asks other members of the household what they would like to eat. Here, Maiken describes the process of 'auditing' her family's meal preferences

> If, for example, there is one day missing [in what to cook for dinner], then they can come up with bids for this. And they have greater influence at the weekend, and I do accommodate them in what I choose, because I know what the family likes.

Participants explained how other family members—including quite young children—are included in shopping as well as cooking activities:

> Well, also because Ane is often sitting on the kitchen table and participating. Also in cutting cucumbers and carrots and yes, depending upon what she is capable of partaking in. She helps grating. You know, when I stand there, but she is allowed to press the button, so it comes out, right. (Jytte)

Adaptation and negotiation is an ongoing process by cooks who are responsive to suggestions and demands, as well as the potential for some responsibilities to be undertaken by other household members, particularly adults. With this type of cooking, time is secondary to the interaction between family members. Here, for example, Marie emphasises family togetherness in the process of meal preparation:

> I think it's … cosy! I love to stand here, and that they bother to come and … you know, a bit like today. You've seen roughly how … It's okay. They have been 50, no 40% more part of it than they usually are. But that thing about how they can go to and fro … or they sit doing something else … or they sit and fiddle with something on their chair and make a drawing.

Among some participants, cooking procedures are explicitly linked with engagement in the sensuous and social pleasures of cooking and eating. Here, for example, Carsten describes including his two sons in cooking:

> I like to de-stress with cooking … This is my space. It's okay that the kids join in, you know. 'Cause if they like it, and I happen to know that Wally likes touching meat and such, and then he shall bloody well touch it, right. And Erik sometimes comes and asks if he can be part of it, and then he is allowed to cut some things, or whatever we are supposed to do.

One of the female participants underlined the pleasures of eating related to cooking: 'But I just love food. So, it's an experience because it's tasty. You know, it's important to us that it tastes nice and that it's a mixture of different things' (Hanne). In the audit approach, cooking procedures can also be integrated with teaching children how to cook, which is then a part of the interaction around cooking and eating in the family. Henrik enacts such an engagement, describing his experience of including his sons in the family cooking:

> To us, it's sort of social with the boys and such, but I also think that you give the children a gift to bring with them by making things from scratch. And it makes me happy that they cook for friends all from scratch without getting told, this is something I've always been preoccupied with that you ought to do. Instead of buying a can of meatballs in curry, it doesn't take longer time to make it yourself.

Betinna draws more strongly on normative values about 'proper' cooking:

> I think it's wrong not to teach your kids to cook proper food. Cooking from scratch. Because you don't know what's in otherwise.

These three different ways of cooking identified among our participants are not case- or country-specific. Indeed, within the existing literature, parallels are evident. For example, an earlier qualitative study of how Danish women in three age-groups cooked (Halkier 2009) suggested a six-fold typology of performing cooking, including planned and improvised types. Likewise, in the UK, Short (2006) highlights timing, planning, organising and being creative as important dimensions of cooking. In a Canadian study, Simmons and Chapman (2012) found cooking to be—among other things—about control over food provisioning and eating, as well as social interactions between family members. Additionally, a time-use study from Belgium (Daniels et al. 2012) found that patterns of cooking could either be described as chore, care or pleasure, all of which are evident in the ways of cooking we identify here.

Cooking with a Meal-Box Scheme

When looking at the ways in which the Danish participants cook when they use meal-box schemes, it is clear that, while this may vary over time as household circumstances change, the dominant way of cooking is the planned way. The meal-boxes make possible particular interactions between materialities, activities, procedures and engagements.

The planned way of cooking demands planning, structuring and timing meals and food provisioning for a longer period, not just for today or tomorrow. This requires allocating and distributing specific parts of the practice of cooking in time-slots, making use of time as a resource and handling how cooking practices are part of ordering temporalities in the family (Southerton 2012, pp. 242–244, see also Chap. 5). The material arrangement of the meal-box scheme seemingly facilitates these processes. Here is an example of a typical narrative about the use of meal-box schemes in planning ahead:

> But when we are sitting down on Sunday, and there are suggestions, and then I usually know what's in the freezer, and they are told, 'no, you can't have meatballs in curry', or it gets written on the shopping list, depending upon what's in the freezer. And then we can go into the app and see what will be coming in the box and plan, so OK, we would like this on the Tuesday, or we would like this on the Saturday or something, whatever fits. And then we make … while we are making this meal plan, we make a shopping list and find out whether we have the things in the cupboards and drawers, and then I usually shop on Monday. (Betinna)

Here, Marie explains how, in her household, they attempt to establish who will be at home when planning ahead with the meal boxes:

> We try to arrange one week at a time with a starting-point in the boxes, which we buy three or four days, depending upon whether there is something exciting. And then we try, the days where most of us are at home at the same time, then we get the meal-boxes.

In this context, the differences between the meal-box schemes are important. 'The Seasons' scheme involves the company deciding the content of the whole box with different options available (e.g. quick box, family box, vegetarian box etc.) with the option to specify the number of people. In contrast, the 'Fairly Easy' scheme allows the customer to decide what dishes to include in their box. These can be selected from a list of dishes for each week; again for a specified number of people. Consequently, Betinna describes planning with fixed boxes, whereas Marie reports planning with adaptable boxes. Thilde, whose household subscribes to the 'Fairly Easy' adaptable box scheme, describes how planning and improvisation is possible:

> You know, typically then … You know I get a mail … something with now this and that has been chosen and such. Then I go in and check if it is the right choices and adapt. Actually, I adapt pretty often.

Once the meal-boxes have been delivered, there is less planning and fewer chores on a daily basis. Echoing earlier findings (Giard 1998; Short 2006), not having to make daily decisions about what to eat relieves a significant burden, particularly for women:

> I think it makes the biggest difference for me, that the daily planning of the meals, plus the time it takes to shop, that this is completely out of the picture. And that's a huge help in everyday life. (Rikke)

Agnes draws attention to the challenges of storage and finding ingredients that might be dispersed around the kitchen which the meal-box scheme alleviates:

> And you don't have to go and find it, you know, you take the box in and up on a table, if you haven't already got it splintered into atoms by which

things should go in the fridge and which things should go into the freezer. Then these things that you need are in the box.

The actual cooking itself—transforming ingredients into a meal—tends to take place in a relatively structured manner because the process involves following step-by-step recipes utilising exact measured ingredients for the specified number of diners, thus eliminating guess-work and reducing the risk of failure or left-over food which might be wasted. Carsten, explains:

> You know, it's incredibly easy to make. If you can just be bothered to actually read the recipe, then it can't go wrong, right?

Some participants highlight this directed kind of cooking as enabling them to serve meals on time compared with those prepared without the use of meal-boxes:

> You know, I think they are really good, because they are fast. Not many of them take more than half an hour … it's very, you know, 'then you do this, then you do that, and then you do this', and after twenty minutes, you're done. You know, if you make something yourself, I can sometimes, oh now they are finished, but they are not really finished. Maybe get it planned or, now I'm soon 42, I can plan how to cook, but sometimes it does slide a little for me, if you get started on the meatballs too late, or whoops, now the sauce has burned, and has to be re-done. (Anja)

However, even when cooking in such an organised and planned manner with the use of the meal-box schemes, multiple elements of adaptation, negotiation and improvisation take place in the cooking. Seen from a practice-theoretical perspective, this is only to be expected, since the reproduction of stable practices through routines involves elements of adaptation, improvisation and negotiation (Warde 2005). Seen in relation to cooking, the existing research also points to the importance of improvisation, for example the 'mothers of invention' in a UK study (Carrigan and Szmigin 2006) who handle different meal provisioning situations in everyday life through juggling not only different foodstuff, but also through adapting and improvising the organisation of the juggling.

While planning clearly dominates in the use of meal-box schemes, it was equally clear that within households that utilised them, dishes also had to be negotiated, adapted and improvised according to variable family schedules. Much of this occurs as part of the cooking process (cf. Halkier 2009; Moisio et al. 2004; Short 2006) and the dominant factor influencing the improvisation of recipes is the tastes and preferences of family members, usually children. As Hanne explains:

> Yesterday … we had that thing from the meal-box, but then I changed it a little bit. It was brisket with squash, carrots and tomatoes, and then with macaroni on the side. And then it actually said, it had to be raw … Then I thought that [son], he is not so super-enthusiastic about the raw stuff, so we just threw everything together in the pan and then mixed it, and put the brisket in. Then it became more ratatouille-ish, and then brisket and macaroni.

In this example, it is the method of serving the ingredients that is changed—from raw to cooked—to make the texture and taste of the vegetables more agreeable to the child. This is reiterated by Carsten, who identifies ingredients which may or may not be 'child-friendly', therefore requiring replacement on the plate:

> Yesterday, we had some oven-baked salmon with some soy-honey marinade on top. And wild rice and artichokes, roasted on the pan with a little lemon and spinach. And then we always make some more greens, because the spinach wasn't so children-friendly yesterday.

Changes in the quantity of an ingredient and changes to a recipe are highlighted by Maria:

> There was this example with putting aubergines in. Where the recipe said half an aubergine with meat-filling, where I thought … I will probably find it difficult to serve this in a persuasive manner. So I tried to make moussaka instead.

Like Carsten's replacement of spinach with a more 'child-friendly' green vegetable, in the exchange below Helen reveals how requested replacements for particular—disliked—ingredients have been normalised in her household's meal-box orders:

> You know, celery, we have been known to take that out, and fennel, we've also taken out. It's been such specific things, that we didn't like. But where the dish could be eaten without it being in there … And then we have saved some of the recipes and then crossed out and written 'no celery', and then instead what else you could put in, such as onions or leeks or whatever.

Supplementing with additional ingredients typically relates to concerns about whether all family members will be sufficiently satisfied at the end of the meal. As Jytte explains:

> I might do a rice-porridge, or I can figure out, if it's some meat, I can go down and buy, I will just buy extra meat—and supplement with vegetables, because I always have those.

There is also the addition of bread, which seems to work as a flexible buffer in such situations:

> Now, they only have boxes for two or four people, and we are five. So depending upon what it is, it might sometimes be that … even though the portions are big, well, if it's something the boys like. They are really capable of eating some meat. So maybe you say, there is only one of these for each, and then you supplement with some bread or something. (Agnes)

Finally, there is also evidence of improvisation where the entire dish is changed completely:

> Sometimes I can sort of just feel that, particularly today, that recipe looks a little bit, what should I say … cumbersome, I just can't manage that today, and then I take, for example, only the meat, and then I use it for something different. (Thilde)

Cooking in Other Contexts

If we look at data from our other work packages and national contexts, it is apparent that the way participants engaged with food varied considerably according to the demands of everyday life in which shopping, cooking and eating are embedded, meaning that different approaches to cooking were combined within the same household. In the UK, for example, Gloria filmed herself cooking meals for her husband and adult son from scratch on some evenings, pointing out that she always made her own short-crust pastry. On other evenings, she used more convenient ingredients such as frozen puff pastry, asserting: 'I never mess about making my own puff pastry; it takes hours and I an't got the time'. Likewise, Tony used convenience products such as a powdered stroganoff mix to add flavouring to a dish made with otherwise 'fresh' ingredients. This, he says, is 'the magical bit'. Tony describes this practice as 'sort of cooking something as opposed to just putting something in the microwave'. However, he did also use a microwavable rice accompaniment for the stroganoff. Reflecting on this in a subsequent interview, Tony reframes the utilisation of such items as a means of looking after himself since they avoid putting him 'under huge stress' but 'in a way that's satisfying to me'. He acknowledges that it is 'very lazy [and] it might seem a bit pathetic' but he insists:

> the extra 10–15 minutes it would take to boil the rice, that is caring for me really, by not giving myself the additional burden, y'know what I mean? Making it into a bit of a treat in a way … 'cause it makes it easier and more comfortable and less of a chore.

'Outsourcing' the hot meal of the day to a workplace canteen follows a similar logic. During the week, canteen customers need not bother about shopping for and cooking hot meals when lunch can be taken very conveniently at the canteen. Two groups of German interviewees described how having a full meal in the canteen fitted very nicely with their everyday practices: people who live alone and do not want the trouble of preparing a full meal for just one person, and people whose partners and children have a hot meal away from home. The time and effort saved during the week pay off at the weekend, when circumstances allow participants to prepare an elaborate Saturday breakfast or Sunday roast.

The data about baby food in Sweden also confirm that different ways of providing meals existed alongside and mixed with each other as the following example shows:

> Most of the time I take a portion from the freezer/fridge and heat it in the microwave. But if I have to cook something, I try to do something quick, perhaps cooked root vegetables or broccoli that I mix with a hand blender together with crushed nuts or tofu. (Susanne)

Other participants such as Jenny confided that she 'makes a lot of food at one time, freezes it and then thaws it in the oven or in the microwave'. Usually she makes two 'dishes of the week' and then varies them. Being able to cook was important but not *the* most important element in the choice of ready-made or home-cooked food as is sometimes argued, although such skills evidently help if you want to cook from fresh ingredients, as evidenced in the quotations above. Just as in the planned way of cooking, planning was important in relation to providing meals for babies:

> I usually plan quite far ahead and gladly prepare something that I can freeze 6–7 servings of. In this way I reduce the risk of 'the child is hungry NOW and I have nothing ready to give her—panic'. When I shop (usually at ICA) I always look to buy the ingredients that I know work to make baby food. Root vegetables, lentils, beans, crushed tomatoes, bananas, avocados, among other things. So I usually do not shop with a recipe in mind, but make sure I have most of what Sally eats at home so that it is possible to quickly and easily throw something together. (Susanne)

This variety of approaches to improvisation and adaptation, even within rather structured ways of cooking, suggests that a range of know-how and skills are a condition for being able to do so. This is the focus of the following section.

Know-How and Cooking Skills

As discussed in the introduction to this chapter, cooking skills include a broader range of understandings, know-hows and abilities to carry out practices and procedures than is conventionally acknowledged. Consequently, there has been a call for conceptually broadening what is understood when referring to cooking skills (Halkier 2009; Lang and Caraher 2001; Lyon et al. 2003; Meah and Watson 2011; O'Brien et al. 2015; Short 2006).

The necessity for a broader conception of cooking know-how and skills is particularly clear when looking at cooking with the meal-box schemes. This includes a reconfiguring of the material arrangements associated with the meal-box schemes (e.g. the use of the internet as an integral part of planning and shopping), different kinds of overlap between cooking and other practices (e.g. everyday logistics and parenting) and explicit societal discourses regarding the normative elements of food consumption as part of the organization of food provisioning (e.g. sustainability and health).

There are several ways in which researchers might contribute to a clarified and extended understanding of the knowledge and skills deployed in food provisioning. First, cooking know-how can be deployed in the kinds of dishes people make and account for when describing their culinary activities. Across the 13 participating families in the Danish case-study, around 45 specific dishes were either cooked in the company of the researcher or referred to during subsequent interviews. Cooking knowledge could narrowly be defined in terms of the number of dishes that are part of a cook's repertoire. Among these households, participants cooked and described traditionally Danish dishes (meatballs, *medister*-sausage, hash, fried pork belly, rice-porridge and rye bread with cold cuts and spreads), those of American origin (burgers, spareribs, steaks and pulled pork), dishes originating in Asia (spring rolls and sushi), Italy (pizza, risotto, pasta bolognaise and lasagne), Mexico (tortillas and frittatas) and Greece (moussaka and saganaki cheese). This variation includes different types of ingredients and cooking techniques. Additionally, other variations were visible in how participants spoke about the dishes they cook, including referring to non-specific dishes, such as soup or oven-roast chicken, or those in which more detail about the components of the dish were listed (e.g. salad with red grape and ginger dressing, burger with goat-meat and figs). The ability (or willingness) to list individual meal components does not, however, indicate levels of cooking ability, knowledge or experience.

A second way of trying to 'unpack' cooking knowledge could be to look at what the participants consider to be simple and complicated cooking. In the Danish case-study, this distinction between simple and complicated cooking seems to be linked with several different specific cooking activities. The difference between more simple meals, cooked on an everyday basis, and more elaborate and complicated meals, cooked at the weekend or for guests, is well-established in the social scientific food literature (Murcott 1983; DeVault 1991; Short 2006). This distinction is also evident when cooking with or without the meal-boxes:

> At the weekend, I make a much bigger effort with … the dinner. You know, I actually really like to cook, and I like to make sushi, but I would never dream of attempting such a sushi adventure here, when I come home from work. (Thilde)

Another classic distinction is put forward by Henrik who juxtaposes cold dishes as more simple, and warm dishes as more complicated:

> Johan and Erik [Henrik's sons], they don't eat rye bread, and not at all rye bread with cuts and spreads. That can be a challenge. In the weekends and the holidays, we often make something warm instead.

A further way the participants distinguish between more or less simple or complicated ways of cooking is directly related to the different boxes available in the meal-box scheme. Here, Hanne explains how different boxes from 'The Seasons' are not as simple to cook, partly because they involve so much preparation:

> When I take the low-carb box or the quick box, I don't think it's easy. They may well say that the quick box is easy to do, but I think there is a lot to prepare. And that's the same with the vegetarian box. You know, so much you have to handle and rinse and chop and what-not.

Finally, the relationship between simplicity and complexity might also be related to knowing how to improvise, an element in cooking already discussed in the section on cooking with the meal-boxes. The ability to understand how to respond and improvise in shifting circumstances, sometimes in the moment, is a skill or competence in itself (cf. Giard 1998; Short 2006). Thilde explains that the inclusion of 'exotic' vegetables in the boxes is sometimes a challenge, but this can be resolved by experimenting:

> There are often rather exotic vegetables. For example, there has been this long period with fennel, right? Where … often actually I think they can be difficult to get used, if we did not have the rabbits. Or otherwise, you have to be really good at making risotto, you know, because everything goes into a risotto in a blended version, right?

Thirdly, when examining how Danish participants cook, both with and without the meal-box schemes, a varied range of cooking activities and procedures seem to be required. Some activities and procedures are shared across the Danish sample. These include planning and organising, shopping, keeping an overview of stock, scheduling alongside other responsibilities and obligations, following recipes, being able to improvise, using left-overs, clearing up and disposal. However, one dimension which seems to be particularly salient when cooking with the meal-box schemes is knowing how to incorporate a wider range of practical concerns in cooking. Across all other differences in the sample, the participants explain how providing 'proper' food is part of the appeal of meal-box schemes, and that through the material arrangement offered by the schemes, they are able to achieve a healthy and varied diet while simultaneously addressing concerns about the environment, food waste and animal welfare. Arguably, this may be interpreted as a valuable cooking skill, prevalent among these (predominantly) middle-class consumers as cooking on a budget is among less affluent households.

Furthermore, a variety of specific types of know-how were performed and described by the different participants. Some belonged more to the planning and organising part of cooking. These were bulk-shopping and bulk-cooking for the freezer, catering for different tastes and preferences, managing the division of household labour around food provisioning, adapting the digital ordering of meal-boxes, coordinating shopping in relation to stocks via mobile phone applications, and sorting and storing ingredients once they had reached the household. Some kinds of know-how were more specific to the process of turning ingredients into meals. Again, knowledge of and ability to accommodate individual tastes and preferences is significant since this not only contributes to successfully feeding the family, but also to the unnecessary waste of food. Marie's description of improvising with aubergines for one of the meal-box dishes (described above) is one such example, where she alters the recipe to make the inclusion of aubergines more appealing to her children who dislike chunky vegetables.

Other kinds of know-how were also evident in the Danish and UK studies, including knowledge of how to assemble a meal using a combination of prepared and raw ingredients, creating new dishes from left-overs, learning to use new utensils or new ingredients, or using ingredients in a different way. Of these, assembling know-how seems to be particularly important to subscribers of the meal-box schemes due to the sometimes unforeseeable demands of everyday life in a family where parents both worked full-time, with dependent children living at home. In these households, participants often ended up re-scheduling planned meals and having to deal, pragmatically, with the consequences of this for cooking. Gro describes a typical situation:

> It was a bit like … you know. It was, what kind of a dish is now left from the meal-box from last week. And then it probably needs to be topped up with something. Oh well, we had left-overs from bolognaise from Sunday. But of course also … I am not fanatical about food waste, but I actually prefer to use the foodstuff I have in my fridge.

A final observation—pertinent across all the Danish households—is what they reported about the sources of their knowledge and skills concerning cooking. Everybody mentioned parts of their social network and the internet. But variations in knowing how to cook were linked with personal and socially-shared concrete know-how, procedures and guidelines and with mediatised know-how, procedures and guidelines as shown in previous research on cooking (Halkier 2009; Leer and Povlsen 2016; Moisio et al. 2004; Short 2006). Participants also explained about learning to cook through their upbringing, where family experiences dominated but also cooking classes at school. Mediatised knowledge about cooking was not limited to digital media but also included cookbooks, magazines and television cooking programmes.

Since meal-box schemes lend themselves to various forms of directed cooking, it is worth discussing whether the material and discursive arrangements of the two different schemes are inscribed with different assumptions about necessary cooking skills. On the basis of a textual and visual analysis of the websites of the two schemes, 'The Seasons' and 'Fairly Easy' seem to share at least two assumptions about the cooking skills required to use them. Both assume that recipes can be followed, and both assume an ability to plan ahead as a precondition for cooking. However, the 'Fairly Easy' (non-organic) scheme—which offers an element of choice—underlines the ability to adjust and adapt flexibly in cooking with the meal-boxes, important dimensions if one wishes to involve children in cooking. In contrast, 'The Seasons' makes a greater assumption about the necessary skills required to manage the inspiration for what to cook.

However, when comparing 'The Seasons' subscribers with those of the 'Fairly Easy' scheme, participants across both highlighted much the same challenges in relation to demands on their cooking knowledge and experience. Among these was how to handle the quantity of food in the boxes, which did not always correspond with the number of people they needed to feed. Here, Hanne expresses this, when comparing two of the boxes offered by 'The Seasons'. One provides too much, the other not enough:

> What is a bit funny is that, if you take 'favourites' [name of the box] for three, then there is an incredible amount of fill [carbohydrate], so we can't always get to use all of that. But if you for example take the 'quick box', then it can seem to be a little bit just below what's needed.

Other issues included how to manage unfamiliar ingredients and new recipes which were sometimes very different from what participants were used to cooking. This was clearly expressed by one of the subscribers to the 'Fairly Easy' scheme: 'You know, all the dishes are nearly never quite ordinary, because there is always just a little twist somewhere' (Carsten).

The 'Conveniencization' of Cooking

Since meal-box schemes are categorised as a way of making food provisioning more convenient (see Chap. 3), degrees and types of convenientness have been discussed within this chapter. However, comparing meal-box schemes with ready meals, processed baby food and eating in workplace canteens, meal-boxes emerge on the lower end of the continuum of convenience. Indeed, the use of meal-box schemes is perhaps the only one of our four cases where participants are required to do more than acquiring, unpacking, heating and assembling. How, where and when can cooking with the use of meal-box schemes be regarded as convenient?

Where cooking can be more convenient in everyday life is primarily at home in the family for evening meals. However, since cooking with meal-boxes involves quite a lot of planning, adaptation and assembling, this kind of cooking can also become convenient in relation to other types of meals, predominantly lunch, where leftovers or deliberate surplus-cooking come in handy.

When can cooking with meal-box schemes become convenient? This can occur in relation to a range of different times-uses (Southerton 2012). Earlier in the chapter it was argued that when participants shifted during the week from cooking 'on their own' without the meal-boxes to cooking with them, the planned way of cooking became the predominant mode of cooking. Thus, if and when planned food-work in the family takes place, cooking becomes convenient by using the meal-box schemes as a planning tool (cf. Hertz and Halkier 2017). This applies not only to planning the content of meals. It also covers the planning of how much time is going to be spent on cooking-related activities, including the shifting of time-uses between planning, meal preparation and other activities such as paid work, clearing up, helping with homework and bringing children to and from leisure activities (as discussed in Chap. 5 on the temporalities of convenience food).

Shopping can also become more convenient by using the meal-box schemes in so far as participants avoid spending time travelling to the shops, using the webpages of the box-schemes on the internet instead, where opening-times are 24 hours and the choice among boxes and dishes is limited—rather than more limited opening hours and potentially endless choice in supermarkets. At the same time, this makes shopping—as part of the preparation for cooking—more flexible in terms of scheduling, something mentioned across differences by the participants.

Finally, both when preparing dinner and clearing up afterwards, food practices can become more convenient by using the meal-box schemes. As mentioned before, cooking with the boxes can be seen as directed cooking, and in this way it can at least potentially function as simplifying family cooking practices. However, this is not always the case as the different adaptations, improvisations and challenges to cooking know-how described earlier in this chapter showed. Clearing up is not very explicitly described among the participants, but the meal-boxes with their measured and weighed amounts and recipes adapted for a particular number of persons are seen as contributing to the reduction of food waste (see also Chap. 9). This might be seen as a parallel to the conclusion in recent research about the amount of food waste being less when using ready-made elements for meals than when cooking meals from scratch (Evans and Welsh 2013). Cooking with meal-boxes can be seen as being a 'ready-measured' kind of cooking.

How can cooking become convenient on the basis of meal-box schemes? Some of the ways have already been suggested. Planning and shopping via the internet are the two main ways in which meal-box schemes become convenient in the acquisition of foodstuff for cooking. In the appropriation part of food consumption, directed and 'ready-measured' cooking seems to be the most prominent way in which cooking with the meal-boxes can become convenient. This also includes the built-in variation of dishes in the material arrangement of meal-box schemes. As parents who cook on the basis of a meal-box scheme, their use automatically brings about culinary variation in meal preparation.

However, it can be suggested that in the appreciation part of this kind of food consumption there are at least three main normative ways of conveniencizing cooking. As discussed in Chap. 4 on the normalization of convenience food, normativity is a combination of what is regularly expected (normal) in cooking and what is considered morally acceptable or legitimate (Halkier 2017a). The first has to do with the built-in variation of dishes in the material arrangements which the Danish participants appreciated since eating a variety of dishes is understood as a standard element of eating 'properly', referred to as part of different food cultural repertoires—not just the official nutritional advice. The second normative way of conveniencizing is the 'outsourcing' of responsibility for the legitimacy of one's cooking on a number of parameters such as health, ecology, animal welfare and food waste. Across all other differences among the Danish participants, this appreciation of meal-box consumption looms large. To quote Gro, who subscribes to 'The Seasons':

> The good consciousness there is about me knowing that someone has reflected upon this, and that it's OK. And that it's organic and that it's good and that it's green, and yes, that it's OK.

The final and perhaps the most important normative way of making cooking more convenient on the basis of using meal-box schemes seems to be that participants are able to claim that they are cooking from scratch and that the family is eating home-made meals, although provisioned in a more convenient manner than 'ordinary' cooking (Hertz and Halkier 2017). Cooking so-called home-made meals from scratch still seems to be the most 'proper' (normative) way of feeding the family.

Conclusion

We have described how cooking is carried out in everyday life, based primarily on the Danish case-study of meal-box schemes. When families cook on the basis of the meal-boxes, cooking performances predominantly turn into a planned way of cooking, compared to when not using meal-boxes, where the participants enact both an improvised way and an audit way of cooking, alongside the planned way of cooking. However, even in a planned mode of cooking, lots of adaptation, improvisation and negotiation take place. The empirical patterns of cooking also support a broadening of the concept of cooking skills. Our evidence particularly underlines the skill of improvisation in everyday life, and the skill of handling societal claims about cooking what is considered 'proper food'.

Finally, the chapter categorised the way that meal-box schemes can be said to become more convenient, even in comparison with other forms of convenient meal provisioning such as commercial baby food, ready meals, and canteen food. Meal-boxes become convenient in everyday life mostly through the differently organised and time-shifting process of acquisition, through directed and 'ready-measured' cooking, and by making normative justifications around 'proper dinners' less socially complicated.

Notes

1. The report can be accessed here: https://www.leeds.ac.uk/news/article/3890/jamie_olivers_cooking_courses_and_improving_eating_habits (accessed 9 November 2017).
2. The perceived benefits of 'cooking from scratch' can be interpreted as part of a wider discourse of nutritional advice, the political implications of which are discussed by DuPuis (2015).
3. *Contra* the discourse of relentless decline, baking has undergone a popular revival in the UK, following the success of TV programmes such as *The Great British Bake Off* (first broadcast in August 2010)—a cooking show that has attracted huge audiences and spawned numerous spin-offs. A similar process can be observed elsewhere in Europe including Denmark and Sweden.
4. The focus group material is not used in this chapter.
5. In Denmark, the main meal is usually taken in the evening. Other patterns apply elsewhere as discussed in relation to our study of workplace canteens in Germany.

9

Convenience, Sustainability and Health

Introduction

Sustainability is a notoriously difficult word to define, often reduced to platitudes about meeting the needs of the present-day without compromising the lives of future generations. In literal terms, sustainability refers to the capacity to persist or to recover after unexpected shocks, sometimes referred to as resilience. In everyday use, however, the term's meanings are now so diffuse and appropriated for so many purposes that it circulates as 'a generally well-meaning but vague notion', lacking in precise focus (Watson 2013, p. 210). Sustainability is a multi-dimensional concept whereby it is acknowledged that environmental conservation may have social or economic costs while the pursuit of political objectives such as economic growth may have damaging environmental consequences. Trade-offs and compromises between competing objectives are therefore a core issue in the pursuit of sustainable development and methods such as triple-bottom-line accounting have been developed to try to capture the diverse (social, environmental and financial) aspects of the concept. Methodologies such as carbon trading have also been proposed to express environ-

P. Jackson et al., *Reframing Convenience Food*,
https://doi.org/10.1007/978-3-319-78151-8_9

mental concerns in economic terms, 'valuing the environment' or undermining its irreducible nature according to one's point of view.

'Healthy eating' is an equally challenging term to define whether one takes a narrowly nutritional view, measured in terms of carbohydrates, fats, proteins, vitamins and minerals, or a less specific approach involving exhortations to consume a 'balanced diet' or to eat 'proper meals'. Government guidance, through ideas such as the Eat-well plate (in the UK) or the food pyramid (in the US), remains ambiguous on critical issues such as whether or not meat should be included as part of a healthy balanced diet. Meanwhile, healthy eating campaigns are often couched in a didactic tone, as in the UK's official encouragement to 'Change 4 Life' by combining dietary improvements with increased exercise. Advice on healthy eating also comes from celebrity sources with limited medical authority, such as Jamie Oliver's Ministry of Food, or in the form of homespun wisdom such as Michael Pollan's invocation to 'Eat food. Not too much. Mostly plants' (Pollan 2009). Assessments of what constitutes a healthy diet are hampered by the fact that the consequences of what we eat may take years to have an appreciable effect on health, with many intervening variables complicating a simple diagnosis or causal explanation. Even in areas such as obesity where the links between body-weight and non-communicable diseases such as Type-2 diabetes, coronary heart disease and hypertension are now well-established (Foresight 2007), controversies persist including the validity of simple measures such as the Body Mass Index or the contested notion of Health at Every Size.[1] These debates also raise questions about whether health interventions should focus primarily on physical health or whether to include mental health and wellbeing (as signalled by the prevalence of eating disorders associated with questions of body-image and fat-shaming).

Even if definitional disputes about environmental sustainability and dietary health were resolved, the connections between them would remain problematic. There have been many attempts to establish the health benefits of sustainable modes of production such as organic agriculture. But claims by accrediting organizations such as the Soil Association about the health and safety benefits of organic food have been disputed. In the UK, for example, the Food Standards Agency contested the scientific evidence on which the Soil Association's health claims were made ('Organic food has no added nutritional benefit, says Food Standards Agency', *Daily Telegraph*, 29 July 2009). The issue is of more than academic interest as there would be clear benefit to environmental campaigners if they could demonstrate that what is good for the environment (in terms of sustainability) was also good for human health—but there is as yet no scientific consensus in this area (cf. Tilman and Clark 2014). Defra's Sustainable Consumption Report also criticises the conflation of sustainable diets and sustainable environments, referring to the difficulties in agreeing what constitutes a sustainable healthy diet, what such diets might look like in practice, and how far they align with current eating patterns (2013, p. 14).[2]

As was argued in Chap. 1, convenience foods are frequently criticised as unhealthy and environmentally unsustainable, citing evidence from the *British Medical Journal* about the nutritional deficiencies of ready meals (Howard et al. 2012) and their resource-intensive nature, responsible for high levels of GHG emissions, heavy demands for energy, land and water resources, and high transport costs (Defra 2012). Others have connected the consumption of convenience foods with increasing obesity rates (and their associated health costs), wasteful packaging and excessive 'food miles' (and associated environmental damage). In a Belgian study (where the market penetration of ready meals is estimated to be over 80%), Geeroms et al. (2008) explored the link between health-related motivation and ready meal consumption. Based on an on-line survey of just under 2000 respondents, the researchers used cluster analysis and analysis of variance to identify five consumer segments, labelled energetic experimenters, harmonious enjoyers, normative carers, conscious experts, and rationalists. Attitudes to ready meals were generally rather negative among all groups, with energetic experimenters and conscious experts showing significantly more positive attitudes. Commenting on the nutritional value of convenience food, Ahlgren et al. (2004, p. 160) suggested that ready meals are usually found to contain too much fat but too little energy to meet the nutritional needs of healthy adults. These characteristics help explain why their survey respondents held negative beliefs about those who were frequent consumers of ready meals. The adverse dietary consequences of convenience food were also explored by Dixon et al.

(2006) who made a direct link between the consumption of such foods and increased rates of obesity and over-weight (see also Alkerwi et al. 2015).

In this final empirical chapter, we consider what our research participants have to say about the health and sustainability of various kinds of convenience food and how the evidence from our case studies might enable us to reframe these issues in new ways. We begin with some examples from the ready-meal study before broadening the discussion, using material from the other three cases concerning commercial baby food, workplace canteens and meal-box schemes. The chapter makes two main arguments regarding alternatives to the standard 'deficit approach' employed in much dietary advice also questioning popular understandings of the environmental impacts of contemporary agri-food systems, as measured in terms of 'food miles' and food waste.

Countering a Deficit Approach

A common assumption in much of the academic and policy literature is that consumers have a poor knowledge of the health implications of their diet and even less knowledge of the environmental costs associated with intensive agri-food production. In an evidence review for the UK Food Standards Agency's *Our Food Future* project, Andrew Darnton argued that 'The most influential factors in food choice are price, quality and taste. Health concerns are subordinate to these, and environmental considerations are lesser still' (FSA 2016, p. 5). Summarising the evidence from recent public dialogues on the food system, Darnton reported that 'When dialogue participants are first told about the links between food and environmental impacts ... their first response is one of surprise' (ibid., p. 6). Survey research leads to similar conclusions. In one study undertaken by the market research company TNS (reported in FSA 2016), consumers were asked to rank the factors that influenced their food purchases. Their unprompted choices ranked price as most important (mentioned by 51% of respondents), followed by quality of food (39%), special offers (17%) and brands (13%). The healthiness of food was ranked lower (12%) and environmental factors such as food miles (4%), non-recyclable packaging (3%) and sustainability of food sources (2%) lower still. Issues such as free-range (8%) and organic production (5%) which could be motivated by health, environmental or ethical concerns were mid-ranked rather than priority concerns. Apart from those with a specific commitment to vegetarian or vegan diets, consumers are very unlikely to acknowledge the environmental impacts of eating meat (cf. Macdiarmid et al. 2016), described by Darnton as a 'lack of awareness bordering on denial' (FSA 2016, p. 38).

In this area, as in many other areas of consumer research, there remains a tendency to blame consumers for issues that are largely beyond their control, to misunderstand the logic that underpins their practices and to address issues at an individual rather than a collective level, assuming a lack of knowledge or a deficit in understanding as the key to desirable 'behaviour change' (Halkier 2017b). Such deficit thinking can be countered by an assets-based approach which seeks to explore the reasoning behind apparently irrational behaviour (such as wasting food that is still perfectly edible or eating food that is known to be unhealthy). Assets-based approaches seek to avoid blame and apportion responsibility in an appropriate manner, challenging the simplistic thinking that lies behind individualistic behaviour change models (cf. Evans 2011; Halkier and Jensen 2011; Meah 2014; Shove 2010). Let us look, then, at what our participants have to say about these contentious issues.

Laura (a retired professional) and other women in the ready meals study were attracted to a 'healthy eating' agenda through its links to dieting and weight reduction. Asked why she switched from cheap white bread as her children grew older, she replies: 'How did I make the shift? [pause] I do think that dieting [pause], weight is probably a major thing, trying not to eat food that's got empty calories I suppose'. Laura has done both the Slimming World and GI diets[3] in the past, explaining that:

> I always seemed to be half a stone heavier that I wanted to be, so I went to Slimming World. I suppose that was quite a healthy diet really, it was a particular arrangement, but the food they suggested, the weighing of them all, it was quite clear that [pause] y'know, it made you aware of how calorific processed, unhealthy foods are … It just crept up on me, this idea of food generating a feeling of well-being. What I have been doing, I don't know, maybe longer than I think, certainly last year, maybe more.

Laura goes on to talk about the GI diet and cutting out wheat, realising that it was making her constipated. She also reports on how the use of certain convenience products such as a jar of artichoke hearts or sun-dried tomatoes make her daily lunchtime salad more interesting, as lettuce, tomato and cucumber can get a bit tedious: 'they are conveniences in that they're not gonna go off and you can just add them to spice something up and make it a bit more interesting'.

Other participants, such as Rageh, who was living on virtually no income as an asylum-seeker, found it very hard to make healthy food choices. As discussed in Chap. 7, his diet of frozen chicken, rice and onions and his preference for instant mashed potatoes was not driven by a deficit of knowledge about what constitutes a healthy diet, but by necessity in terms of what was most affordable. He understood the importance of diet and exercise but did not have the energy to engage in physical exercise because he was almost permanently hungry.

Tameka was also clear about the benefits of 'healthy eating': 'I try to be healthy, yeah'. She spoke of her preference for eating fruit in between meals or, if not, offering her daughter, Graça, a multigrain savoury snack which she described as 'slightly healthier crisps'. When asked where her concern for health came from, she replied: 'I'm trying to lose weight [laughs]. I try to eat "clean", as I call it'. But, she says:

> I'm not always committed. I like a pizza sometimes [laughs] and say on a Saturday, when I've been doing a spring clean or whatever, by the end of the day I'm like, I had intentions to cook but I forgot to defrost the meat or something silly like that [so she ends up getting a Chinese take-away or a pizza]. I'm like, 'Graça, what do you want for dinner?' 'Pizza!'

Tameka describes her Friday night take-out phenomenon when she gets home from work and can't be bothered to cook. She also explains how what she would like to cook is sometimes thwarted by not having time or money to do a 'proper shop' having 'stretched it too long and missed the boat' (when she had her wages), sometimes having to pay for her shopping with her state benefit (working tax credit). Again, there is no deficit of culinary knowledge in this case but circumstances (including finances and the preferences of other family members) prevent her shopping, cooking and eating as she would like.

Other participants, such as Tony, are perfectly aware of the discourse of 'healthy eating' but rebel against it for various reasons. He knows what and how he is supposed to eat and that he drinks and smokes too much. Deploying a kind of hierarchy of convenience, he notes that buying convenience foods to heat up one-self is a step up from buying take-away food. His eating habits are closely related to the changes in his personal life, having recently split up with his wife and moved out of the marital home. He gestures towards more healthful foods by eating cereals such as Weetabix and porridge pots or what he calls 'nicer versions' of the blander 'cardboard' alternatives. He says: 'I find muesli a bit sort of boring [pause] and a bit sort of self-punishing, so we have the treat-y [sic] equivalent of it basically'.[4] He also explains how eating single-portion ready meals such as cottage pie and lasagne prevent him for eating larger quantities of food as he used do when he lived with his wife.

Tony also talked about some salmon which he described as 'a concession to being healthy', pointing out that it was quite easy to cook and tended to last a reasonably long time. Tony's attitude to 'healthy eating' is expressed in a rather arch and ironic way, drawing attention to his taste for convenience foods such as cup-a-soups, Pot Noodles and ready meals. This was also apparent when describing a late-night cooking session when he rustled up 'a disgusting pile of slop [laughs] which I really like doing actually'. The meal consisted of two cans of soup, retrieved from his 'cupboard of shame', combined with a pack of instant noodles, a sachet of flavouring and some grated cheese. Tony explains that these late-night cooking sessions were something he used to do a lot when he was married, creating a private space for himself where he could eat 'elementally psychological, comfort food … gooey … a very oral experience', acknowledging that there was 'something almost foetal about [pause] curling up on the couch with the telly on, spooning this gunge into my mouth'. While he is fully aware that such comfort food is unhealthy, he clearly finds it comforting, adding temporarily at least to his sense of wellbeing.[5]

Tony is only semi-apologetic about cooking such 'disgusting' food, talking reflexively about how problematic meals had become during the last year living with his wife and anorexic daughter. Now, he enjoys shopping for himself and describes Pot Noodles as 'my secret vice'. Having described how he had eaten a ready meal on his own on Christmas Day, he insists that it was from Tesco's Finest range and 'was actually very nice'. He also uses self-deprecating humour to deal with the perceived shame that attaches to being overweight:

> But what I do quite a lot, I think, is about the shame thing, is make a, a bit like everything else, like the overweight thing, is turn it into a joke, self-deprecating joke, so I do draw attention sometimes, it's a big joke [among his colleagues], me and my Pot Noodles.

Again, he is fully aware of the discourse of 'healthy eating' (and there is no knowledge deficit to be overcome), but he describes his practice as a 'counter-attack', being 'slightly cool' and 'alternative' in what he eats and how he accounts for it. He also laughs about his aversion to taking leftover food to work in a Tupperware container, the complete opposite of Laura and her husband Ted for whom the use of leftover food is considered a virtue.

Other participants also provide a rationale for departing from what they understand to be 'healthy' eating options. In one case, for example, Maryam tries to provide her family with healthy home-cooked meals, made with a combination of fresh and processed ingredients, and her husband, Imran, chooses to eat healthily when at work. But their eldest son rejects her efforts, preferring sugary snacks and greasy take-away food. However, since their doctor has expressed concern about their son being underweight, Imran is prepared to indulge the child:

Maryam:	Every single night I cook. I love it. I love to cook and give the nice food to the family, that's just me in general.
Angela:	*Do they appreciate it?*
Maryam:	No [laughs].
Angela:	*[laughs] I guess your husband does?*
Maryam:	He does, he likes healthy food ... He helps me, like, he cuts the fruits, he warms the milk for the kids, he's really taking care of them, like, he makes sure, in the morning ... he comes down and warms the milk up and says 'This is for you, this is for you, please have it'. [But he] also buys things

> like cheese strings, he knows that it's convenient in the morning, um, at least they had something, like those small yoghurt drinks, yoghurts and the kids like, sometimes, these cupcakes, lots of the different things like, umm, pancakes, already wrapped … They're taking in the car, so these kinds of things, he shops for those things, I don't shop for those things.

Despite the apparent conflict between Maryam and her husband, Imran excuses his behaviour on the basis that it is preferable for his underweight son to eat unhealthy (processed, sugary) foods than to eat nothing at all ('at least he's eating something'). The children take advantage of Imran's leniency, asking him to take them shopping, because, as Maryam says, 'they know they're gonna buy whatever they want' (yoghurt drinks, cheese strings, chocolate, crisps, fancy desserts, crème brulée, cakes and buns)—things which, Maryam says, she really hates and never buys. The additional irony is that Imran teaches on a human nutrition course so his food choices are not born out of any lack of medical knowledge. Another model of comfort and care seems to be operative in this case, where Imran expresses his love for his children by indulging their tastes for food that his wife would not choose.[6] He also bargains with his children, offering them a glass of milk and some chocolate or allowing them to have some crisps if they eat half a chapatti, also taking them to various sports activities at the weekend. In this case, it seems, several competing paradigms of health and fitness are in play, none of which can be reduced to a simple deficit of nutritional knowledge.

Another participant explained his apparent indulgence towards his daughter using a similar logic. James is a single father who travelled a lot with his job and was often away over-night. While he would buy fresh ingredients for his daughter, she would rarely eat them. So he also bought high-end, gluten-free ready meals for her as a way of ensuring 'that she'd always got some meals to look after herself with', recognising that, had he not done this, his daughter—who had coeliac disease—'would just have eaten rubbish'. While James acknowledged that ready meals were not ideal, he felt that they had a place 'when time's against you', offering more variety and being healthier than take-away meals, for example. In James' logic, ready meals provide an acceptable way to feed his daughter while he cannot be physically present to cook for her. James took a similar approach to feeding his father who has dementia, introducing his mother to Wiltshire Farm Foods, a company that delivers frozen ready meals to care homes and housebound elderly or infirm people. Ordering low-fat, low-salt ready meals was, he thought, preferable to the kind of food that his mother cooked for his father, not having taken his diabetes into account. Again, we see competing logics in play rather than a simple lack of dietary knowledge. Similar ideas about competing and overlapping ways of handling societal contestation over mundane consumption practices have been described in research on how people connect with media messages and public campaigns. For example, Halkier (2016) identifies the *empowering* way of handling contestation, where the issue of health or sustainability is already on the popular (kitchen table) agenda and consumers feel supported by the campaigns. Contrary to this, Halkier identifies an *anti-disciplining* way of handling contestation, where campaigns are seen as tools of social control which are rebelled against. And, finally, there is the *pragmatic* way of handling contestation, where bits and pieces of campaign issues become negotiated and fitted-in to mundane activities.

Beyond 'Food Miles' and Food Waste

Returning to the review paper for the UK Food Standards Agency's *Our Food Future* project (FSA 2016), Andrew Darnton identifies two partial exceptions to the assumed ignorance of consumers about the environmental implications of contemporary agri-food systems: 'food miles' and food waste.[7] This is somewhat ironic as the validity of the 'food miles' concept has been questioned in professional circles while it is still prevalent in public dialogue exercises such as those conducted by the Global Food Security programme (GFS 2016) and the Consumers' Association (Which? 2015). Referring to the distance food travels 'from farm to fork', the term has been criticised for over-simplifying the energy demands of imported food which in some cases, such as Spanish tomatoes grown under natural sunlight, may be lower than 'local' tomatoes grown in the

UK with artificial heat in greenhouses and poly-tunnels (cf. Defra 2005). Other issues, associated with food waste, are also more complex than they might at first appear including the amount and type of packaging which consumers often assume to be 'environmentally unfriendly' while industry experts proclaim their environmental benefits, prolonging the shelf-life of food and reducing waste (cf. Stuart 2009). Again, then, let us turn to our case studies, starting with the UK research.

Many of our participants were adamant that they wanted to avoid food waste. Even when they acknowledged that prepared foods, liked stew packs, may not be 'the best way to buy things', they argued that they were beneficial in avoiding or reducing food waste. Consider this exchange between Angela Meah (who conducted the interview) and Deborah and Edward (a working-class couple in their 40s):

Edward: It's like a stew pack, it's fresh veg but it has everything you want in, so there's no waste, you throw it all in, it's in with the stew and it's got everything you need.
Angela: So that's quite important to you then, avoiding waste?
Edward: Absolutely.
Deborah: Yeah, I don't like to waste things.
Angela: So is that about, the whole waste avoidance thing, is that about cost or…
Edward: No, it's disgusting throwing food away when some people are hungry. No, I can't stand it.

Interestingly, Edward's disgust is related to the needs of other people who are going hungry rather than to more abstract concerns about wasted resources or environmental sustainability.

Other UK participants such as Gloria were equally insistent about not wasting anything including some bits of leek that were stuck inside a food mixer: 'you know me, waste not, want not'. She also took some scraps of cabbage to a friend to feed their pets so there was no waste:

Gloria: I don't like waste; I can't stand waste … if something goes off and I have to chuck it away I get really angry because, to me, it's a waste. I don't like wasting food.

Angela: *Why?*
Gloria: It's money, it's wasteful, and plus there's people who would love to eat that and can't. I don't like waste.

Other participants, such as Jem, had a clearer environmental motivation for avoiding or reducing food waste, reporting an aversion to over-packaged products such as the Tesco Finest range of cherry tomatoes which come in a plastic carton, sitting on a piece of Bubblewrap, which Jem thought was unnecessary. He also had a commitment to buying local food:

> Yeah, food miles, I will generally try to buy as local as possible [pause] in vegetables, in fresh stuff, and I get very annoyed with packaging, um, because it just seems farcical [pointing toward the cherry tomatoes on the table and giving the example of buying some nuts for lunch which come in plastic tubs].

Some participants acknowledged that they wasted food because it had already gone beyond the expiry date before they were ready to eat it. But they were apologetic about such behaviour, refuting accusations of consumer profligacy that others have also criticised (cf. Evans 2012). Consider this exchange with Maryam in which she reports a need to 'take care of' food:

Angela: *Um, do you ever find yourself having to throw food away?*
Maryam: Yes.
Angela: *Because either you've not eaten…*
Maryam: No, you're not gonna believe, I just threw [pause] 5, 6 tins yesterday, mushroom soup and the other soup.
Angela: *Why did you throw them away?*
Maryam: Because they were out of date. Because I bring new things and…
Angela: *Oh, so you just put them at the front?*
Maryam: This is my, this is my really bad habit, I don't check that the food is getting out of [date], I don't know, I don't use much of the tinned food to be honest … My husband is really

> conscious about this. I mean I don't mind to eat bread if it's in the fridge and it's two days out of date or three days out of date. I mean I check if it's not, but they NEVER, it's me who is trying to finish, I don't waste, I absolutely hate it, to be honest, because I really like to take care of the food.[8]

Maryam goes on to talk about reheating leftover food:

> Like cooked food, like if I cooked yesterday, straight away I put 'em in a glass [pause] pot and cover it properly with cling film or with a lid and keep them in the fridge, and I know this food is good for another day, or two days even … Sometimes it's left over, I keep that in the fridge and I only warm the section which I'm gonna eat, so I eat, even the third day, the fresh food made but I don't want to waste that, and even the chapattis, sometimes you cook and…

Angela asks if Maryam makes her own chapattis and Maryam replies:

> I made my own dough and my own chapatti every single day, so even that chapatti I keep 'em in the fridge and I warm it up the second day instead of throwing [it out] and, for example, if the food is STILL there and I could not eat, I throw to the birds, the chapattis or the rice.

Maryam describes wasting food as 'a bad habit' that she tries to avoid, countering the assumption that consumers just don't care about food waste. Interestingly, too, Maryam seems more reluctant to waste food she has cooked herself than tinned food that she has bought, suggesting that home-cooked food has higher value, reflecting the labour and care she has taken in cooking it (cf. Moisio et al. 2004). There are some exceptions, however, such as her use of frozen sweetcorn which she justifies on the grounds that whole corn on the cob is very expensive and might go to waste. She concludes this exchange with a discussion of the need for 'balance' and her concerns about the chemicals in processed foods:

Maryam: So I think sometimes convenient food is really good but it depends, you have to have a balance, you don't have to have

> everything, for example, you're buying ready meals and you just warm it up. You don't know what's inside, who made it and how many chemicals and how processed it's been.

Maryam raises another question of 'balance' later in the interview, concerning the way some convenience foods might reduce the amount of food waste when cooking for just one person:

> Any extreme is really bad, some people just buy the tinned food or the ready meals and they're not really bothered to cook and, on the other hand, if you need just one cup of soup and you're making all that effort, buying all the ingredients, maybe you're wasting half of them because you don't need that much, so it, it should be, um, in the middle somewhere.

Several participants spoke about the way their good intentions regarding the avoidance of food waste were thwarted by the exigencies of everyday life, recalling David Evans' analysis of the many everyday occurrences that lead to unintentional food waste (Evans 2014). For example, Tameka talks about how frozen vegetables might help her to avoid buying bags of onions and then throwing them away half-used when they start to go off: 'especially if I mean to cook and I don't end up cooking as often as I would, you know'. The subsequent discussion is full of moralized language about the gap between her actual practice and the ideal to which she aspires: 'Yeah, it really hurts me when I waste vegetables and cucumbers and stuff, but … I stopped buying cucumber 'cause we never used it how we should'.

Some households in the current study have been involved in our other research projects including an earlier study of consumer anxieties about food (CONANX).[9] This has enabled us to develop a longitudinal element to our work, with the researcher (Angela Meah) working with the same participants in some cases for up to 7 years, as well as comparing their views with members of different generations of the same family. In one family, for example, food waste emerged as a particularly salient concern among retirees, Ted and Laura. Born within memory of the Second World War, the couple shared a concern with the avoidance of food waste, a concept that Laura spontaneously raised as a food 'anxiety' when she was first interviewed: 'we haven't talked about wasting food much, but I mean that's like a thread right through from, you know, being a kid after the war, you just didn't waste anything'.

During one of many cooking observations with the household, Angela mentions how Ted and Laura's son—who is less concerned with food waste—had nonetheless reported that he will save bits of leftover vegetables (including 'a few bits of cabbage in a bowl') to pass on to his parents (see Watson and Meah 2013, p. 110). Reflecting on her mother-in-law's frugality, Laura responds:

Laura: We'll eat anything [laughs] and we'll be cross if we find out it's been thrown away … I can hear your mother saying in her sort of Welsh way, 'I hate waste, I hate waste'. I can't quite do it, did you ever hear her say that?

Ted: Yeh, 'oh I 'ate waste, oh dear'.

Angela: But was that partly because she had so many mouths to feed on so little money?

Ted: Yes, that and also when she was growing up, y'know, in the '20s there wasn't much around unless you were well-off, so you really had to make things last,

Laura: Mm.

Ted: I mean, because everyone has got parents and grandparents like that but unless you have [stutters], the reason she did it was because she had to stretch the food out, but the reason that someone like I do it is, is because I think that this stuff is precious, shouldn't be thrown away really, and grown and nurtured and cooked.

Ted's belief in the 'preciousness' of food is reflected in his attitude to the issue of 'food miles', where he insists that 'food should be eaten at source, when it's in season and ready to eat':

Ted: There's plenty of stuff to eat, you don't have to bring things from Israel and Peru and Australia to eat. There's lots of food around, local food, and if you're not in the tomato season, eat something else.

Angela: *So, are food miles, is that a concern for you?*
Ted: Yeah … I think it's disgraceful.
Angela: *Is that about like, carbon footprints?*
Ted: No … I wouldn't put in carbon footprints, that's a modern way of saying that but, it's a waste of world resources to bring food, perishable food, a long way in an expensive aeroplane to sell at inflated prices to people that can't afford it.

Continuing this argument, Ted talks about the versatility of tinned tomatoes ('they're great in tins and you can do great things with them'). While 'fresh tomatoes are wonderful' in season, Ted compares them to 'the ones you get from Tesco in February' which have little or no taste. There are trade-offs, then, between seasonality and taste, where (certain kinds of) convenience food such as tinned tomatoes meet Ted's needs without incurring unnecessary environmental costs (which he describes here in terms of 'air miles').

In contrast with her husband's emphasis on the global management of food resources, Laura is primarily concerned with what occurs at a household level. While her anxieties initially focus on fish 'laying in the fridge for more than a couple of days', she also spoke of her evolving concerns about a range of products which have undergone 'preservation' processes to extend their consumable life. While Ted reports having been happy to use a jar of shop-bought pickle that had been languishing in his (now deceased) father-in-law's larder for almost two decades, Laura is more cautious about jars of sun-dried tomatoes and pickled cucumbers which have not been 'used up' before they start to show signs of visible deterioration.

Laura: One thing that worries me are products that say 'once opened use within…' and it's stored in the fridge. I sometimes rake around in the fridge and find things that look pretty sorry for themselves … particularly on the pickly type of shelf. [Reporting on a jar of (sun-dried) tomatoes that caught her eye in the fridge since coming back from a recent break, she continues:] They were deep in oil but I was sure they had like grey mould or something, they seemed to have a sort of greyness around them and I pondered it for a bit and then I just threw it all away.

Angela: *Would you [Ted] have still used it?*

Ted: Mm, no, not if it looks horrible, all sorts of things can happen ... There might be some impurity in, in, in the process of putting it in the jar which has caused that, you see use-by on jars of olives, we keep them in our fridge for months.

Laura: I wouldn't worry about them, but I mean one thing that we have, that I've been concerned about is buying those big jars of pickled cucumbers, like sandwich slices, and that we haven't always kept them in the fridge, and I found them in the cupboard and the liquid was sort of moving about and, that's not right, that should go, I mean we do now, we do keep them in the fridge.

These comments show that Ted and Laura think carefully about food hygiene and its implications in terms of food waste, where some preserved (convenience) foods offer an extended shelf life providing they are stored correctly once opened.

Reframing Health and Sustainability

The other case studies also contribute to our reframing of the health and sustainability of convenience food. In the case of baby food in Sweden, for example, environmental sustainability was primarily addressed in reference to eco-labelled and locally-produced products. Parents describe how they opt for eco baby food, often in line with an overall effort to buy eco-labelled food products for the whole family:

Helene: *Do you usually opt for a specific brand of these jars?*

Karoline: Yes, I usually buy Hipp's ecological cans

Helene: *Yes, OK. So you take note of these things?*

Karoline: Yes, I want to buy ecological products.

Helene: OK, so you told me you try to avoid salt when you cook yourself. And you try to mind the iron level. You choose products that are iron fortified. Are these the two things you prioritize?

Karoline: Yes, primarily.

Helene: So nutrients but also this ecological bit?

Karoline: Yes, and that applies to the whole family.

Helene: So this is something you prioritize in your household?

Karoline: Yes, as much as possible. Obviously you cannot buy everything eco-labelled but I try to find as much as possible.

For Karoline and other parents in the Swedish study, eco-labelled baby food products are a convenient way to fulfil an overall desire to cook and eat eco-labelled products. Eco-products also meet desirable health criteria ('nutrients but also … ecological'). Most of our participants used eco-labelled and non-eco-labelled products interchangeably as they do with 'fresh' and 'convenience' foods. A minority distinguished between fresh ingredients and processed or convenience products. For example, Ebba describes how she tries to buy eco-labelled fresh fruits but does not really involve eco-labelling in the choice of processed food products or baby food, while Frida describes how she always opts for eco-labelled baby food jars and pouches but chooses a non-eco-labelled porridge since the only brand available is rice-based and makes her baby daughter constipated.

Parents who buy eco-labelled products offer various reasons for doing so. Besides a general awareness of the unsustainability of contemporary food production and an effort to care for the environment, parents also describe the baby's preferences as central to them choosing eco-labelled baby food:

Maria: So when you buy food, do you usually buy a particular brand or do you alter?

Jeanette: Well, we have noticed that there is one brand that she prefers. In the beginning we tried several brands but then we noticed that there was this brand that she liked more than the others. And now we only buy this brand.

Maria: *What brand is this?*
Jeanette: Hipp.
Maria: *OK it's Hipp. Yes, many babies like that. It seems as if they make tasty food.*
Jeanette: Yes, I am not sure what it is about that [brand]. But the first six-months, then it was very, it was proper puree while the others had pieces. Maybe that was the thing. But now when there are pieces in Hipp's products as well it is still that one she prefers.
Maria: *So you started out trying different brands and products and then her preference guided your choice. So Hipp being eco-labelled is not something…?*
Jeanette: No, but I see that as a bonus. I try to buy eco-products at home. However, not only eco-products. So I see that as a bonus.

Some parents, like Jeanette above, link the eco-label to taste and caring for the baby's preferences, regarding sustainability as a desirable bonus. Similarly, Sara is very careful about what she eats. Away from home, she maintains a vegan diet as a way to handle the insecurities associated with commercial food production. At home, when she is able to make sure that products are eco-labelled and soundly produced (local, free range), she occasionally eats meat and cheese. She describes eco-baby food (and ready-made vegetarian substitute products) as a convenient way to offer her baby safe food that is produced in a sustainable manner.

Many parents compare and contrast eco-labelled products with locally-grown products. Some, like Anna, are very careful to buy Swedish products:

Anna: We only buy foods produced in Sweden. It has to be [Swedish].
Maria: *OK. What about eco-labelled food?*
Anna: No, that is where I draw the line.
Maria: *So it has to be produced in Sweden then?*
Anna: Yes, well we have started to look for eco-labelled products. But you know. When the eco-bananas cost two SEK more, I

can handle that. But paprika, one cost 20 SEK and the other 49, well then … No, we do not eat that much paprika. And I know a farmer that produces crops according to eco standards but he cannot afford all these certificates. He said we do not use pesticides and we do all the things you are supposed to…. So we, I do not trust that [eco-labelling].

Ulrika also expresses ambivalence in regards to eco-labelled products. She finds it hard to know whether these products are better than others:

Ulrika: I do think about this but it is so damn hard knowing what to believe when it comes to ecological foods. You know, you really want to feed him the very best quality food. And then you think it is too expensive, and I think—how bad is the regular food and how ecological is the eco-labelled food? How different are these products? How much does it increase the risk of various diseases? There is still so much you do not know. Is it really that bad? That is where I stand. Maybe it is better to avoid processed foods and skip that bit. And just try to cook yourself. It is really hard to know. You know they use a lot of crap in the food.

Here, it seems, Ulrika is unwilling to trust commercial baby-food producers to label food accurately. Nor is she prepared to pay the premium that such foods command without a clearer understanding of the risks involved.

Parents who opt for eco-products often describe how they weigh the eco-label against locally-produced products:

Henrik: We buy eco-products when we are able to. But I always prioritize locally-produced products before eco-labelled ones. I think that local producers, they cannot afford to piffle with these things. If you sell your crops to the people you live next door with you cannot afford [to produce bad food]…

Maria: What about baby food? Is that something you look for when you buy food jars, that it is locally produced or eco-labelled?

Henrik: Many of them are. It says so on the can. Many of them are eco-labelled. And that is really good of course. It is very comfortable/comforting [*skönt/översättning*]. There should not be any sugar or added ingredients. She only gets good quality food.

Parents described products from the Semper brand as particularly trustworthy. Semper is the market leader in Sweden and its long history is probably central to the parents' understanding of the brand as 'Swedish'. Many parents also seem to think that Semper's products are produced in Sweden. Some of them are and Semper porridge and *välling* are marketed as produced in Sweden. One mother, for example, buys only Semper *välling* because it is locally produced in a small town close to Falköping. However, as Sara noted, Semper do not produce their food jars in Sweden and the ingredients are not necessarily of Swedish origin. This is, however, not something that the company communicates to their customers. Trust and transparency are definitely key issues here.

While most parents commented on eco-labelling and locally-produced products, they seemed less reflective when it came to waste. Food waste is an issue addressed by some of the parents and weaning is described as a wasteful practice in the sense that babies might not eat all the food prepared for them. They might eat some but also just chew and spit out or play with the food. Some parents described preparing home-cooked taste portions that they kept in the freezer. Annika is the only parent to address the discrepancy between the size of the food jars and the amount of food consumed by babies:

Annika: So I try to buy ecological baby food. Semper's eco-labelled products or Hipp.
Maria: That is important to you?
Annika: Yes, there have been exceptions, but I try at least.
Maria: And then (as you mentioned) they do not eat that much from each jar. At least not in the beginning.
Annika: No, there is a lot of waste.
Maria: What do you think about that?

Annika: Well, I do not want to waste food so if there is a lot left in a jar I have fed her some and then I've put the rest in the freezer in small cupcake moulds to bring out little by little.
Maria: So, you thought about sustainability in that manner as well?
Annika: Yes, I try not to waste.

Again, our evidence points to consumers being concerned about environmental issues and adjusting their practices to minimise unavoidable waste.

Packaging is another issue that is often addressed in relation to food and sustainability. It is clear that parents value the convenience that come with commercial baby food and that the packaging is central to the way weaning is carried out. But they also take account of the way everyday life with a baby is shaped and organised. As discussed in Chap. 5, packaged food helps parents manage the sometimes demanding temporality of baby feeding, allowing them to be more mobile. In this sense, packaging is also central to the spatiality of weaning and baby-care (discussed in Chap. 6).

It is also interesting to note that when we asked parents about packaging they brought out its positive aspects. When issues of sustainability and packaging were addressed, parents were ambivalent about plastic packaging while glass jars were described as less problematic:

Helene: So, you have not cooked for her yourself then?
Felicia: We tried to but she would not eat it. But then we are not that good at cooking for ourselves either. Ready-made is rather convenient.
Helene: What about all the jars and packaging?
Felicia: Well, I figure it is recycled. And I recycle some when I make jam and so on. But otherwise we hand it in to the glass recycling.
Helene: So, that is no problem for you then?
Felicia: No. In that case, these Ella's Kitchen products are more of a problem since they are plastic. That feels a bit troublesome. Glass jars feel more OK.

This view of glass packaging as unproblematic or even a means for acting sustainably can be related to the work of Gay Hawkins (2011, 2012) on food packaging as central to the production of the identity of 'the recycler' and the political as well as economic performativity of food products and packaging.

The workplace canteen study also raises various sustainability issues particularly concerning food waste. One canteen (at the rehabilitation centre) aims to reduce waste by combining side dishes and main dishes. This provides customers with a bigger choice and less likelihood that unwanted food will be wasted. They also allow customers to have a second helping so there is no need for guests to overload their plates on the first visit. Food waste also occurs in the kitchen as well as in the dining area. Canteen kitchens have to calculate quantities very tightly but do not want to end the service with no food left so some waste is inevitable. Hospital canteens have to provide enough food for everybody though they may not know in the morning exactly how many people they have to cater for later in the day. Canteen managers try to avoid wasting food because it is economically sensible and some offer left-over food to staff at the end of the day.

Canteens might be considered to contribute to sustainable consumption in a wider sense since they reduce the energy and transport costs of individual consumers shopping and preparing meals for themselves. They may also have specific 'environmentally friendly' policies reducing the waste associated with packaged food, plastic coffee cups and take-away food containers. Taking account of these issues would involve a wider environmental 'audit' of different modes of consumption but were not often discussed by participants in our research.

Finally, the meal-box project also raised some important issues regarding sustainability and health, particularly the organic scheme 'The Seasons'. During our conversations with participants it became clear that striking an appropriate balance between cost and quality was a critical issue. Countless differentiations were made between different kinds of organic food, some stressing the specific properties of particular foods. For example, Helene claimed that she could only eat organic nuts because of food allergies, while others claimed that organic oranges and lemons had additional health benefits compared to non-organic fruit.

Some participants claimed that there were marginal differences between organic and non-organic food products in relation to animal welfare, while others thought that organic chicken was too expensive and that corn-fed chicken was equally tasty. Taste was also relevant in the case of organic red meat where participants talked about the difference in taste between organic and non-organically produced meat. Many participants talked about other families' food choices in relation to organic food, claiming that other parents, particularly mothers, were hysterically committed to organic food, to the point where they had to hide the fact that they bought non-organic pizza. Overall, however, most participants focused on their everyday lives where more abstract issues such as environmental sustainability were less prominent (perhaps also due to them having somewhat 'outsourced' these issues).

While relatively few participants made a connection between organic food and larger environmental issues, the main exception was the desire to avoid food waste, reusing rather than throwing out left-over food. According to Jytte:

> It's very modern that we shall spend less on food … or have less food waste. And of course we must do that. So you should stop buying more than you need. For us, I think that we do it as well as we can, exactly because we feed our leftovers to the rabbits and the hens. We have waste-sorting and we are actually quite good at that. But it's not as if we have a lot of it … but that's again because our rabbits they just eat [laughs], and our hens they don't do so much, they are too small … And I'm not in doubt about there having been much too much food waste and what's in the shops today. Terribly much food waste. And I think it's really good that they have made all those things in there in Copenhagen, where they drive around and give all the goods to social institutions … It makes good sense, because it's too stupid just to throw it out.

Some participants, such as Hanne, mentioned the few occasions on which they wasted food, claiming that there was very little that they throw out. The exceptions seem to involve ingredients that they dislike and had not ordered, where they tried to improvise with other alternatives:

> Mushrooms [*østershatte*] which are really, what do you call it … rotten. Where what was supposed to be on the top of the dish or something, and you just think, bugger, that taste is sort of missing, or tomatoes where something has happened, or a salad which is completely soft. Where you sit and, like, OK, we didn't get the salad on the side then. Then you sit with a lot of potatoes or rice without anything on the side, and then something's sort of missing … Or we are not so keen on celery, so if there's a celery thing, then we sit sort of like and … or if there is something we already have in the fridge instead.

Other participants also improvised in order to avoid wasting food. Henrik, for example, described how, if they were cooking something on Thursday evening and had one dish left over, he would always swop things around to fit what they had. Maiken talked about how days 'moved around', while Jytte described how they try to figure something out, choosing a meal where they have most of the ingredients already rather than having to go out and buy everything from scratch. Others talked about supplementing what was already in the fridge to avoid throwing food away which can lead to a 'hotch-potch' by the end of the week.

Health and environmental issues were only sometimes linked in accounting for their choice of organic or non-organic food, including those with specific allergies. According to Anja, for example:

> If I buy a chicken, I buy an ordinary one, luckily you can get them everywhere now, you know fresh chicken. But I don't think that it has to be primarily organic or something else. I want to have it organic, and especially organic fruit and vegetables, and that's because I am birch-allergic, and there are many cross-reactions with apples and such. There you can clearly feel the difference between organic and non-organic, with how much you react. And that's because there are pesticide residues and such in the peel that you react to. 'Cause I can easily go out and eat an apple from our own garden, without my mouth beginning to itch, but I simply cannot sink my teeth into a non-organic apple.

In other cases, participants were content to leave environmental issues for their suppliers to resolve. As Rikke argued, for example:

> You know, I'm happy that it's organic and I am happy that inside of that you can choose local. I realized that in the 'Fairly Easy' [meal-box scheme] there is also often something from Spain and what-not, but there is still a focus on it being food that's in season. And it's not ... you know, I don't go much into the details, but when I just skim over things, as I have done before, then I just think, well, OK, I can be calm about this, you know ... absolutely calm. Then I don't have to think. I can just think that it's all right and it's not sprayed and there are a lot of things that I agree with them about, also animal welfare and such. And I don't have to know exactly where these pigs come from, because I have ... you know, they've sold the idea to me so well, that I know they are in fucking control. And I don't have to go into more detail.

In this case, 'Fairly Easy' relieves Rikke of the need to make difficult ethical decisions including trade-offs between organic and imported ingredients or between animal welfare, pesticide residues and provenance (cf. Halkier 1999; Meah and Watson 2013).

Finally, the use of meal-boxes can reduce food waste by providing the exact quantities of food needed for each meal. Anja felt she made too much food before getting the boxes delivered, while others talked about how they used to have 'cupboards filled with all sorts of stuff'. Hanne suggests that it is mainly wrong deliveries that cause food waste, including food that they didn't order and don't eat.

Conclusion

In this final empirical chapter, we have sought to challenge the common assumption that all kinds of convenience foods are equally unhealthy and unsustainable. We have shown that consumers use convenience foods (often in combination with other kinds of food) for a wide range of reasons: to provide emotional comfort, to meet the dietary needs and contrasting food preferences of different family members, to fit in with the temporal demands of daily life, and to provide variety and choice. The use of convenience foods, in our empirical material, was rarely fuelled by a lack of dietary knowledge or culinary skill suggesting that deficit models are an inadequate way of framing the subject. Rather, we suggest, an assets-based approach, rooted in an ethnographically-informed understanding of the practical logics which underpin what is commonly seen as 'consumer choice', might be a preferable alternative.

Similarly, we have argued that food is rarely wasted because consumers simply do not care about it. On the contrary, we have seen how food gets configured as waste as a result of the exigencies of everyday life and that it occurs despite, rather than because of, the way food is valued by consumers. Our research suggests that we need a better understanding of the way that 'food choice' is embedded in everyday social practices which render some kinds of food more convenient than others. In normative terms, making healthier and more sustainable practices *more convenient* might be preferable to blaming consumers for their apparent ignorance and presumed profligacy.

Finally, we have shown that consumers can make a connection between food and environment, both in relation to the avoidance or reduction of food waste and in relation to sustainability, which they frequently cast in terms of 'food miles'. While we acknowledge that 'food miles' is of questionable value as a measure of environmental sustainability, this should not be taken to reflect a lack of concern on consumers' behalf but as highlighting the need for a wider 'accounting' of the environmental costs of different kinds of food including social and cultural issues as well as narrow economic measures of environmental impacts.

Notes

1. Health at Every Size is a social movement whose purpose is to encourage bodily acceptance and self-confidence, often by the rejection or questioning of dieting advice (https://healthateverysize.org.uk, accessed 21 December 2017). For criticisms of BMI and related measures, including compulsory weight checks for school children, see Evans and Colls (2009). For a wider commentary on the politics of obesity, see Guthman (2011).
2. The link between healthy and sustainable diets is not expressed so equivocally elsewhere in Europe. The Danish Consumer Agency has, for example, previously made a positive connection between organic and healthy food, including practical advice for consumers.

3. GI diets aim to reduce the glycaemic index by cutting out starchy foods.
4. The use of treats and rewards (as discussed in Chap. 5) is a common practice among our participants, corresponding to the culinary antinomy of health and indulgence described by Warde (1997).
5. Thanks to Hannah Baker, a Masters student at the University of Sheffield, for making this valuable distinction between comforting and comfort food.
6. See Meah and Jackson (2017) for a wider discussion of the dynamics of convenience and care.
7. Consumers are also concerned about other environmental issues such as pesticide residues although their level of knowledge is often criticised, focusing on those who are suspicious of 'chemicals in food'. Again, there are important differences across Europe, including Denmark, where a high level of knowledge about environmental issues is more normalised.
8. On the common confusion between use-by, sell-by and best-before dates and their implications for food waste, see Milne (2013).
9. The CONANX study was funded by the European Research Council through an Advanced Investigator grant awarded to Peter Jackson. Angela Meah and Matt Watson were responsible for the work package with UK consumers.

10

Conclusions

The FOCAS project on which this book is based (addressing the relationships between Food, Convenience and Sustainability) aimed to answer four questions: how 'convenience food' is understood by consumers and how its use relates to understandings of 'healthy eating' and environmental sustainability; with what specific practices (shopping, cooking, eating, disposing) convenience food is associated; how such foods are incorporated within different household contexts and domestic routines; and to what extent current practices are subject to change (towards more sustainable and healthier practices). We begin by reviewing some of our key findings at the general project level before drilling down to more specific conclusions from each of our case studies (on baby food, ready meals, workplace canteens and meal-box schemes). We conclude the chapter by drawing out the implications of our research for policy and practice, ending with some suggestions for future research.

Key Findings

The FOCAS project was funded via the ERA-Net SUSFOOD programme as part of its research strand on understanding consumer behaviour to encourage healthier and more sustainable food choice. Approaching

P. Jackson et al., *Reframing Convenience Food*,
https://doi.org/10.1007/978-3-319-78151-8_10

the subject counter-intuitively, we focused on a category of food that is often considered to be unhealthy and environmentally unsustainable. The category of 'convenience food' eludes easy definition (as discussed in Chap. 3) and, we argue, can be thought of as a chaotic concept that combines things that are logically distinct and divides things that are conceptually similar (cf. Sayer 1992). Rather than trying to identify the common properties of all the different foods that are arbitrarily described under the label 'convenience food', we examined four types of food that might all be considered 'convenient' though in very different ways. Our case studies included commercial baby food, supermarket ready meals, workplace canteen food, and home-delivered meal-boxes. We approached the research from a comparative perspective with case studies from Denmark, Germany, Sweden and the UK. Although we focused on one category of food in each country (meal-boxes in Denmark, canteens in Germany, baby food in Sweden and ready meals in the UK), each case study included at least some comparative research in one or more of the other countries. This leant the project a critical dimension which, while not a formal case-by-case comparison, allowed us to highlight what was distinctive about each case and each country, including the translational issues that we outlined in Chap. 3. The definitional complexities of convenience food could have been handled in a variety of ways: focusing on specific kinds of convenience food (such as supermarket ready meals) or tracing the way research participants used the term. We used both of these approaches on different occasions and for specific purposes. But our principal method was to trace the way different kinds of food come to be regarded as convenient in particular circumstances, a process we refer to as 'conveniencization' (cf. Halkier 2017a).

Throughout the book we have distinguished between 'convenience' as a means of saving time, making things easier or otherwise facilitating the everyday lives of consumers, and 'convenient' food that enables this process. While 'convenience food' conventionally refers to specific kinds of food such as ready meals or frozen pizza, our research demonstrates that all kinds of food can be 'convenient' in different household situations and in different domestic circumstances. Using left-over food from meals that were cooked from scratch, for example, can be convenient even if the contents would not normally be regarded as 'convenience food'. As well as the analytical manoeuvre of distinguishing *convenient* and *convenience* food, we also deploy the concept of *conveniencization* to describe the practices and processes through which particular kinds of food provide different kinds of convenience for consumers in different circumstances. So, for example, home-delivered meal-box food provides a convenient solution for consumers who wish to avoid meal planning and weekly shopping trips even though most of the content of the boxes would not conventionally be regarded as 'convenience food'. This approach enabled us to escape from the conceptual cul-de-sac of arbitrating which foods to include as 'convenience food' and which to exclude from such a category. Instead of focusing on certain kinds of food as belonging in *a priori* terms to the category of 'convenience food', we focused instead on the practices and processes that are involved in making consumers' lives easier or more convenient.

While 'convenience food' might be a useful term in the context of food marketing or retail category management, our research demonstrates that the term makes little sense at the level of household practice. Far from being a separate category, we found that most households combine different kinds of food and different methods of food preparation to varying degrees in their everyday lives. Conventional oppositions between 'home-made' and 'convenience' food, between 'fresh' and 'processed' ingredients, and between cooking 'from scratch' and more cursory forms of food preparation such as microwave cooking, are not tenable analytically as most households combine these different forms and processes in practice without much conscious reflection. Despite their analytical deficiencies, such distinctions are widely used in everyday life and pose particular problems for mothers who face social disapprobation for their (over-)reliance on convenience foods, given the persistence of gender stereotypes about 'feeding the family' as predominantly women's work (De Vault 1991). These issues were at the heart of Chap. 7 on the moralization of convenience food where our participants frequently used self-deprecating humour and irony to excuse their use of convenience foods and culinary short-cuts. Far from being a self-evident concept, then, 'convenience food' is caught up in all kinds of ideological, moral and ethical debates whose implications we have sought to tease out throughout the book.

Our research supports those who challenge the distinctions between 'traditional', 'home-made' and 'convenience' foods which Grinnell-Wright et al. describe as 'not necessarily helpful' (2013, p. 22). Similarly, Marshall and Bell (2003, p. 62) insist that convenience and home-made foods are part of a continuum, not two separate categories, distinguished by context not content (see also Warde 1999, p. 519). In a seminal paper, Grunert (2003) concluded that 'convenience' is not a clearly defined concept but usually involves making something easier, saving time or mental effort at various phases of meal preparation including planning, preparation, eating and cleaning up afterwards (see also Scholderer and Grunert 2005; Jaeger and Meiselman 2004). Grunert further suggests that convenience foods may be a substitute for meals taken inside or outside the home, with movement possible in both directions (where consumers use convenience food as a cheaper substitute for restaurant meals or when their experience of restaurant eating affects their choice of particular kinds of convenience food). Our research supports all of these suggestions including the convenience of eating when 'on the move'.

The use of convenience food is often described as providing meal solutions for time-pressed consumers, issues we discussed in Chap. 5 on the temporalities of convenience food. These time pressures are associated with the challenges of synchronising domestic schedules within busy modern households. Convenience foods may help in time-saving (reducing the amount of time spent on specific tasks) or time-shifting (enabling consumers to manage their time through practices such as batch-cooking, home-freezing or reheating left-over food). Our participants stored tinned, frozen and other kinds of convenience foods in anticipation of future needs, referring to their 'war-chest' or 'nuclear bunker'. There are also important temporal dimensions to the way different kinds of food, such as processed baby food or supermarket ready meals, have become embedded in our everyday lives, a process we refer to as 'normalization', involving both what is statistically average or commonplace and what is normatively approved or disapproved (as discussed in Chap. 4).

From a 'theories of practice' perspective, we explored how people are recruited to new practices and how, in turn, they may relinquish older practices. We insist that these practices are *social* practices that cannot be understood in terms of individual 'consumer choice'. It is also useful to think about the different time-points at which 'convenience' may be valued when planning a meal, purchasing food, cooking or cleaning up. The canteen and meal-box case studies illustrate this process best, where collective provision at lunchtime may have knock-on effects for meals at other times of the day and where commercial provision of meals with exact quantities and detailed recipes may relieve households of the responsibility for meal planning and shopping while retaining the social approval that is associated with home-cooking.

The increasing use of convenience foods is often linked with the alleged decline in cooking skills and culinary competence where, we argue in Chap. 8, media debate and political rhetoric appears to be running ahead of the evidence. A similar argument can be made about the health and environmental implications of convenience food where our research focuses on the reasons why consumers adopt practices that depart from expert opinion and official advice, rather than taking a 'deficit' approach (assuming a lack of knowledge or skill).

Our research also challenges the common assumption that consumers cannot see a link between food and the environment. As we argue in Chap. 9, consumers often do make such connections, albeit in terms of the academically-contested concept of 'food miles' or in terms of the reduction or elimination of food waste. These are quite narrow terms for engaging with the environmental implications of convenience food compared to the wider framing that would be raised in a more thorough environmental 'audit' of the resource demands associated with all stages of the production and consumption of convenience foods.

In all of these ways, our research suggests the need to re-frame our understanding of convenience food, for both academic and policy-related reasons. We turn now to the findings of each of our four case studies, beginning with the study of processed baby food and everyday feeding practices, before outlining the implications of our work for policy and practice.

Processed Baby Food and Everyday Feeding Practices

Based on fieldwork in Falköping, a small Swedish town, and comparative (desk-based) research in Denmark, Germany and the UK, the research found that the use of commercially processed foods in baby weaning is not simply a matter of choosing one product or one specific cooking technique. What is considered 'convenient' baby food is sometimes home-cooked and sometimes ready-made, depending on a range of factors such as the baby's food preferences, the family's food tastes, daily meal rhythms and ideologies of eating together. Local weaning practices are combinations of materials (bodies, taste buds, jars, spoons, microwaves), competences (cooking, planning, feeding) and meanings (ideas about being a 'good mother', feeding the family well, eating proper family meals). 'Convenience' is an emergent category (comprising specific combinations of meanings, materials and competencies), rather than a property of particular kinds of food.

Our analysis of the marketing strategies of baby food companies challenges the distinction between 'convenience' and 'care', showing that, in some circumstances, convenience food products can carry qualities that are equated with care, as opposed to the conventional idea that 'convenience' products are inherently negative or even harmful (cf. Meah and Jackson 2017). Marketing and other web-based material on baby food and weaning practices includes a range of text and images offering discursive resources that support consumers in negotiating actual and ideal practices. Our work with Swedish mothers shows that the health risks associated with weaning and baby food are major concerns for public health officials, while mothers are more concerned about practical matters to do with baby-care and everyday life. Taste preferences matter most when it comes to choice of baby food. Good nutritional quality and price are also important, but sustainability is not a big concern, although locally produced food is preferred. For the Swedish-born women, in particular, conflicting information on weaning practices from public health officials is a big concern.[1] Environmental issues are mostly confined to eco-labelled products and notions of locally-produced products. There are also trade-offs between price and organic labelling, but the taste preferences of the baby and other everyday concerns matter most.

The Healthiness and Sustainability of Supermarket Ready Meals

The ready meals case study focused on the UK, with comparative work in Germany. These countries comprise two of the three biggest markets for ready meals, the UK accounting for 42% of European sales and Germany for 20%. UK consumption of ready meals is twice as high as France and six times higher than Spain. While 30% of UK adults are reported to eat ready meals more than once a week (MINTEL 2013), the households in our research had much lower levels of ready meal consumption than these national averages suggest.

In contrast to the popular view that sees 'convenience' and 'home-made' food as mutually exclusive categories, our research demonstrates how convenience foods such as ready meals are frequently combined with other kinds of food in participants' everyday domestic practices. Within these combinations of different food categories, a 'hierarchy of acceptability' appears to exist where some kinds of food (such as take-away food or frozen ready meals) are ranked lower in terms of social desirability than other kinds of food (such as pasta dishes and chilled ready meals). Previous research has shown how mothers are able to combine different foods in inventive ways, artfully preserving the labour-saving benefits and 'necessary concessions' associated with convenience food with the caring qualities associated with home-cooking (see, in particular, Carrigan et al. 2006; Carrigan and Szmigin 2006).

The popularity of ready meals can be understood in terms of wider social changes (such as increased female participation in the labour force and increased numbers of single-person households) and how it 'fits in' with people's other commitments and responsibilities. Our research shows how the use of convenience food can be justified in terms of the opportunity it affords to cater for the diverse tastes and culinary preferences of different family members, to provide food at times that suit the disharmonious domestic routines of parents and children, to allow children to cook for themselves when parents are out, to try new kinds of food or to enable more 'quality' time to be spent with family members. Some participants also justified their use of convenience food as a strategy for reducing food waste, buying prepared food in the quantities they required rather than loose, pre-bagged or packaged ingredients that might spoil before use. This was particularly true for older people, those living alone and households without dependent children.

There are some interesting variations in the point at which 'convenience' is sought, whether at the point of purchase, during storage or when cooking or clearing up after a meal. For example, storage seemed to play a greater role in the use of convenience foods in Denmark where consumers valued the convenience aspects of its acquisition and storage (Costa et al. 2007, p. 86), while some UK consumers freeze chilled ready meals for future use but generally buy them for more-or-less immediate consumption (within a day or two). In this case, 'convenience' is located primarily at the point of cooking and eating rather than acquiring or storing food.

We also explored the 'stocks of knowledge' through which participants made sense of their dietary practices. Sources of knowledge include inter-generational transmission (often from mothers and older female relatives) but also food-related media (where tinned tomatoes and frozen peas are presented as equal or better in taste and nutritional value compared to fresh ingredients). While there was some evidence of nostalgia towards the food and cooking skills of previous generations, many participants commented on improvements in cooking and the wider range of food available compared to their parents' or grandparents' generation (over-cooked vegetables being a recurrent theme). Finally, we found that consumer practice was often at variance with expert knowledge and official guidance (such as on 'correct' refrigeration and storage instructions). This suggests a need to understand the everyday logic that informs consumer practice rather than assuming them to be ignorant or careless in their everyday domestic routines.

Canteen Food and Workplace Food Geographies

Canteen food represents a special case in terms of the 'conveniencization' of food where collective provision increases efficiency but may restrict individual choice. Unlike other kinds of convenience food which may be consumed 'on the move' or eaten alone, lunch in a canteen is as much a social event as an opportunity to satisfy appetite. The social aspect of workplace eating should therefore be taken into account in the discourse of healthy eating.

This case study explored the role of workplace canteen food in the daily eating routines of two mid-sized German towns (Düren and

Euskirchen). In these towns, eating routines are still characterised by the traditional hot lunch as the main dish of the day and the smaller evening meal or *Abendbrot* ('evening bread'). Accordingly, workplace canteens primarily offer hot meals. Some divergent trends were noted in terms of the temporal structure of eating practices. A move away from a 'hearty' main dish with meat and side-dishes, rich in carbohydrates, towards lighter and/or vegetarian meals is occurring where women are more likely to have a meat-free dish than men. Simultaneously, meal-times are becoming less predictable with a gradual diversification from the stable pattern of breakfast between 7.00 and 9.00 and lunch between 11.30 and 14.00. Canteens are coping with this trend by providing take-away solutions such as sandwiches beyond the main eating hours. Customers welcome this increased choice and wider availability of different kinds of food but these changes impose more constraints on canteen staff in terms of irregular working hours. In canteens (unlike restaurants), employees' working hours are very structured and increased flexibility for guests demands increased flexibility in the organization of work and the timing of breaks for canteen staff.

Apart from the meal choices made by canteen guests, key decisions are made at earlier points in the supply chain which affect the kind of meals served in canteens at a larger scale. The legislative context is set by EU food safety regulations and HACCP-style risk management systems.[2] Further measures are provided at local or national level including dietary guidance, health and safety regulations and public procurement policies enforced through calls to tender which tend to favour more industrialized large-scale food producers. Food choice also depends on whether a canteen is subsidized or not and what budget is available to be spent on food. Catering businesses operate in a very price-competitive environment.

Customers have variable degrees of freedom to improvise their meal choices, substituting side dishes or main courses, selecting food from the salad bar or buffet, asking for larger or smaller portions, taking food home with them and so on. This is especially evident for vegetarians or people who do not eat pork (the 'standard meat' in German canteens). Caterers are aware of the special dietary needs of their Jewish or Muslim customers and some smaller canteens prepared individual pork-free dishes on demand. While meat dishes were generally the most popular, selling out first, vegetarian dishes were also consumed by non-vegetarians, depending on the kind of meals served. Besides these practical considerations, the canteen study also contributed important findings to Chaps. 5 and 6 on the temporalities and spatialities of convenience food.

Meal-Box Schemes: Convenience and Sustainability

The fieldwork for this case study was undertaken in Denmark with some comparative web-based work in Germany, Sweden and the UK where meal-box schemes have variable market penetration. The case study provides another 'special case' in terms of the reframing of convenience food, moving away from the conventional focus on particular types of food (such as ready meals or take-away food) in order to understand how food provisioning can be made more convenient in terms of everyday family life. The study looked at two home-delivered meal-box schemes where customers are provided with carefully measured ingredients, set recipes and detailed cooking instructions, relieving them of the responsibility for planning meals. One scheme was organic ('The Seasons'), the other non-organic ('Fairly Easy').

Meal box-schemes are a good example of the kind of food provisioning which is experienced as convenient by parents, because the food-work involved in planning, shopping and preparing meals is simplified. Meal-box schemes help to maintain the everyday logistics of family life in various ways. When families subscribe to a meal-box scheme, they switch from a composite way of food provisioning, including elements of planning, improvisation and interaction with other family members, to a highly planned and directed way of providing food and cooking. Traditional convenience practices are used in some of the participating families, but they are associated with morally 'bad' food, lacking freshness, of lower quality, with less healthy properties and likely to be more environmentally damaging compared to 'proper' home cooking. Meal-box schemes function as a convenient way to avoid the disparaging associations of convenience foods, combining the virtues of home-cooking with the convenience of precisely measured ingredients, detailed recipes and cooking instructions, and minimal responsibility for meal planning.

Meal-box food is understood as living up to a number of claims placed on 'proper' food, such as variety, freshness, locality, sufficient to satisfy family members and of good nutritional composition. Across organic and non-organic meal-boxes, parents also understand them as helping to reduce food waste. The food in organic boxes is seen as more sustainable in general, relative to the food in non-organic boxes. Parents negotiate the provision of (what they see as) 'proper food' around discourses of more or less healthy food and more or less caring parental practices, passing over the responsibility for complex ethical and practical questions to the meal-box provider. Providing for, cooking and eating family meals is thus a constant attempt to balance a number of different, sometimes contradictory, expectations about cooking in a sufficiently healthy, varied and loving way.

The institutional arrangements around meal-box schemes uniformly emphasise convenience, where the consumer is constructed as in need of help to make daily life easier. At the same time, there are clear differences between organic meal-boxes which tend to stress sustainability and non-organic meal-boxes which tend to stress nutrition. While environmental issues are important in terms of the content, quality, reduced food waste and provenance of meal-box food, consumers do not seem as concerned about the wider energy or environmental costs of home-delivery (in terms of packaging, refrigeration and transport costs), supporting our earlier argument about the need to reframe the health and sustainability implications of convenience food.

Implications for Policy and Practice

The four case studies all have significant implications in terms of policy and practice. The baby food case suggests that the convenient practices associated with commercial baby food are a prerequisite for managing everyday life for families with small children, but that this need not be opposed to healthy or sustainable food habits. Packaged food helps parents manage the sometimes demanding temporalities associated with baby food and feeding, allowing them to be more mobile. Organic (or ecological) products for babies also need to be reasonably priced and preferably locally produced. Providing healthy, sustainable and affordable foods that are compatible with the mobility and other demands of modern family life is therefore the key challenge for food manufacturers and marketers.

Mothers and babies are not inherently vulnerable to foodborne risks, but sometimes become vulnerable in encounters with specific risks, real or perceived. Risk assessments by public health officials are not identical with the risks perceived by mothers, which means that women are not always given what they regard as relevant advice, and that risks reported by the mothers can be overlooked or ignored by health authorities. Improving the quality and relevance of public health information and providing it in the format and at the time that mothers require it should therefore be a priority.

The ready meal study shows that, in a saturated market, new product development is focusing on innovations such as healthier and more sustainable options as well as premium ranges and products targeted as specific groups such as dieters and children. Some of these developments offer win-win outcomes in terms of commercial, health and environmental benefits and should be encouraged by health authorities.

Consistent with our commitment to a 'theories of practice' approach, our research suggests that change towards more sustainable or healthier modes of consumption should be approached in a systemic way *at the level of practice* rather than in terms of the individuals who carry out those practices. Given the socially embedded nature of food and eating practices, behaviour change initiatives that are cast at the level of individuals are unlikely to succeed. In particular, our evidence suggests that focusing on 'consumer choice' at the individual level is a flawed model for changing food-related practices, given their fundamentally social character.

Dietary change and the adoption of more sustainable culinary practices are likely to occur through a process of gradual accumulation and steady sedimentation rather than through sudden innovation. As outlined in Chap. 4, the 'normalization' of practice has both technical and cultural aspects. Changes need to 'fit' with the circumstances of people's everyday lives and be both culturally acceptable and technically feasible (cf. Halkier 2013). More generally, we advocate an assets-based rather than a deficit-based model of consumer practice, drawing on an ethnographically-informed understanding of the 'stocks of knowledge' that consumers use to make sense of their practices, whether or not these correspond to scientific knowledge, expert advice or official guidance.

The canteen case study demonstrates that the dominant factor in canteen food provisioning is the available budget to spend on food ingredients. Fundamental improvements regarding the quality and diversity of canteen food are linked to an increase in budgetary resources. Subsidies may be justifiable if the health and/or sustainability benefits of collective food provisioning can be demonstrated. More research on the working conditions of canteen staff and the benefits of public procurement policies should also be done.

While meat is likely to remain a popular option in most workplace canteens, vegetarian recipes, suitable for mass catering and meeting the restrictions regarding financial and temporal budgets, the availability of ingredients and cultural conceptions of a 'proper meal' should be promoted. Consideration could also be given to other innovations such as 'meatless Mondays' and the use of meat substitutes that may have dietary benefits and be more sustainable than animal-based protein.

Finally, the meal-box case study highlights the need to focus on the way food provisioning is organised in more systemic terms rather than focusing entirely on single food products. This is an area where more sustainable and healthier food practices can be supported by market-led innovations in food service management. In line with our 'theories of practice' approach, the meal-box case underlines the need for changes towards more sustainable and healthier modes of consumption to focus at the level of food practices rather than in terms of individual behaviour change. For eating is, as Warde (2016) argues, a *compound practice*, embedded within other practices of shopping, cooking and waste disposal.

Meal-box schemes, like workplace canteens, also raise important questions about the nature of choice. Parents who subscribe to the schemes and customers who eat at canteens seem happy to relinquish some choice over what they eat in order to enjoy the 'convenience' of more organised or collective modes of food provision. In these circumstances, they are prepared to trade-off a more limited menu or selection of food in order to avoid meal planning and shopping, and (in the case of canteen food) cooking and washing up.

Our research suggests that everyday life involves a complex range of practices that require scheduling and coordination with other family members and wider institutional constraints (such as inflexible working hours, diverse leisure commitments and multiple travel arrangements). Meal-box schemes appeal to consumers as a way of 'handling' the many tasks and purposes that we confront in our everyday lives, making food and eating more convenient. Unlike other kinds of 'convenience food' that are socially derided and morally disapproved, such as take-away food, meal-boxes are a 'do-able' means of providing healthy and sustainable meals in ways that can be 'fitted in' to the logistical demands of everyday life, providing what are seen as 'proper' (home-cooked) meals without the demands of meal-planning or the need to shop for individual food items. Among our participants, meal-boxes solved practical problems of getting food on the table, while living up to official standards of variety, nutrition and (to a certain extent) reduced environmental impacts. The challenge for policy and practice is to see how this model might be extended and whether it can be made more socially inclusive since the costs, particularly of organic schemes, are relatively high. A wider environmental audit of the costs and benefits of these schemes might also be undertaken in order to calculate whether the reduction of food waste, achieved through the provision of food in the exact quantities required for meal preparation, is offset by the costs of home delivery.

Future Directions

Research on 'healthy eating' and environmental sustainability are areas where political rhetoric, media commentary and public opinion frequently outrun the available evidence, providing an inadequate knowledge base for sound policy development. Two specific gaps have been identified as in need of further research. The first concerns the link between (various kinds of) convenience food and the alleged decline of cooking skills where existing data are incomplete and increasingly outdated. The second concerns the relationship between consumers' cooking and eating practices and a wide range of other social practices within which these specific practices are embedded. The domestic scale is under-explored in terms of the energy, food and water demands of household consumption, although we note some recent innovative work in this area.[3]

Our research also challenges the received wisdom about consumers' alleged inability to make connections between food and health, and even more so between food and the environment. While we have found stronger evidence of consumers' interest in food waste and 'food miles' than in many previous studies, there remains an imperative to situate these specific concerns and dubious measures within a wider understanding of the environmental costs and benefits of our current, heavily industrialised, agri-food system. Principles of environmental audit could be extended to include a wider assessment of the health and sustainability of everyday consumption practices, acknowledging the difficulties of 'valuing nature' without making a reductive calculation or ignoring their social and cultural dimensions. Reconciling the competing claims that are implicit in the concept of 'healthy and sustainable diets' would be an important first step.[4]

Finally, our research has suggested that the moralization of convenience food is a significant stumbling block in understanding the social dynamics of contemporary food consumption. Reframing our understanding of convenience food within the social practices of everyday life is proposed as a viable alternative approach to the current emphasis on individualized behaviour change initiatives. Moving from the conventional deficit approach towards more assets-based and ethnographically-informed understandings of the logic that underpins consumer practice is suggested as a potentially valuable way forward.[5] In sum, our research has led to a reframing of convenience food, challenging the conventional emphasis on its dietary deficiencies and lack of environmental sustainability. Rather than making a blanket condemnation of convenience food as uniformly unhealthy and irredeemably unsustainable, we have taken a practice-based approach to the subject, seeking to learn from its commercial success in meeting consumers' needs and reframing that knowledge in ways that might help identify healthier and more sustainable alternatives.

Notes

1. The study also included a number of Somali-born women, not reported here.
2. HACCP stands for Hazard Analysis and Critical Control Point methods, a standard procedure for ensuring food safety.
3. See, for example, the ESRC-funded 'domestic nexus' research led by Matt Watson at the University of Sheffield with colleagues from the Sustainable Consumption Institute in Manchester (https://nexusathome.wordpress.com/).
4. This is one of the research themes being explored by the Sheffield Sustainable Food Futures research group (SheFF) as part of a wider agenda for integrated, system-wide, interdisciplinary agri-food research (Horton et al. 2017).
5. The case for an assets-based approach is set out in more detail in the final chapter of *Anxious Appetites* (Jackson 2015). A specific application to the question of foodborne disease is explored in Jackson and Meah (2018).

Appendix: Research Participants

The baby food study:

Amanda, 27, works as inventory and order administrator at a trucking company. Her daughter is 8 months old

Anette, 28, has been a project employee of an IT company, carrying out office work, now unemployed. Her daughter is 7 months old and her first child

Cecilia, 33, lives at her in-laws' farm. Her daughter is the youngest of four

Karin, 37, a teacher and mother of nine-month old boy and an older boy and girl

Ella, 21, stopped musical school when she got pregnant, now unemployed. Her 7 months old son is her first child

Felicia, 28, PhD student in biology. Her daughter, 10 months old, is her first child

Ingrid (and **Richard**), 22, with a 7 months old son. Child-minder at a day-care center

Jessica, 32, pre-school teacher, 8 months old daughter with an older brother

Jenny, 30, is a teacher. Her son is 5½ months and her first child

Judith, 22, working at an institution for disabled children. 8 months old child who is a first child

Klara, 31, process engineer, son soon to be one year old and a first child

Katarina, 28, nurse, her seven months old son is her first child

Petra, around 30, is trained and works as a chef, her seven months old son is her second child

Susanne (and **Håkan)** are around 28 and both work as teachers. They have a 9 months old daughter

P. Jackson et al., *Reframing Convenience Food*,
https://doi.org/10.1007/978-3-319-78151-8

Sigrid, 31, works as a manager at a travel agency, her daughter is 7½ months old and her first child
Signe, 30, assembling car engines at the Volvo plant in Skövde. Two sons, 7 weeks and 2½ years old
Ulla, 34, worker at a factory that makes insulation. 9 month old son with an older sister
Veronica, 22 (and **Peter**) working in the food industry, in a factory which produces cheese. The son, eight months, is her first child
The men who are mentioned in brackets participated at the meeting point during the interview

The ready meal study:

UK	Germany
Ken and Val (72), he is a retired professional, she works part-time; homeowners; financially comfortable; rural. Foodwork shared, but undertaken mainly by her. White British	***Thomas, Susanne*** and daughter ***Franziska*** (7): Professionals, university educated, both working full-time and living in Bonn
Edward (48) and ***Deborah*** (46), full-time employed in low-skilled jobs; working class; no children; financially comfortable; homeowners. Foodwork undertaken by her. White British	***Klaus*** (former teacher) and ***Petra*** (former development worker), retired couple without children, have travelled/lived abroad, university educated, living in Oberwinter
Gloria (47) and ***Jack*** (45). Working class; full-time employed; local authority housing; low income; unemployed adult son lives with them. Foodwork shared, but undertaken mainly by her. White British	***Amara, Yarif*** (mid-40s) and two primary school-age sons, Syrian refugees, living with the wife's brothers in third-floor flat in Bonn North, currently receiving state support and living off their savings. Before they fled Syria she used to be a teacher and he was an engineer
Laura (69) and ***Ted*** (71). Middle class; retired professionals; financially very comfortable. Foodwork shared, but cooking done mainly by him. White British	***Carl*** (23) and ***Maria***, student flat-share in Alfter, living off parental support and part-time job
Tameka (25). Working class; works part-time; single mother (one child under 6); local authority housing; income topped up with state benefits. Black British [African/African Caribbean]	***Nora, Ricco and Hannes***, live in a student flat-share in Poppeldorf, all in their final year of their bachelor degree, not receiving student loans

UK	Germany
Jem (31). Middle class; professional; good income but has debts; living in shared privately-rented accommodation; non-resident partner. White British	***Oscar, Linda and Harald***, flat-share in Bonn Endenich with one student, one gardener and one person about to start a MA, aged 25–30
Phil (44). Working class; unemployed; living in a supported community for people with alcohol and substance misuse issues; very low income (from state benefits). Mixed race	***Rudolf***, divorced man with new partner. He lives in Brohl, she lives in Lohmar. Food bank customer, late 50s, skilled worker, injured in road accident in 1983, currently not working
Rageh (30). Seeking asylum in UK; very limited income; living in temporary charitable housing; unable to work; non-resident child under 3. Somali	***Minna*** (59), food bank customer, mother of three grown-up children plus one child she gave birth to when she was 17, given up for adoption against her will. Now separated, lives alone. Used to work as a care nurse, suffers from severe back pain. Very basic education
Maryam (39). Middle class; works part-time; married; 3 children aged 9–15; homeowner; financially comfortable. Foodwork done largely by her (husband does some top-up shopping). Pakistani	***Lena*** (37) and children: Birk (3) and Maya (1.5). Children's father is in prison; she is an ex-drug addict, currently unemployed. Living in a small second-floor flat. Born in Thüringen (East Germany), family fled as soon as the border opened in 1989
Tony (56). Middle class; professional; good income but has debts; recently separated; child under 14 who visits a couple of times a week; private rented accommodation. White British	***Nicol*** (late-30s) and ***Karsten*** (mid-40s), children Felix (11) and Johanna (13). She works for a charity, he is a soldier, currently working in Bonn. Dual income, two cars; consider themselves middle class, university educated
James (49). Married; professional; homeowner; good income; two children aged 15+ from previous marriage (dependent child visits); has indirect caring responsibilities for elderly parents, one of whom has advanced dementia. Foodwork shared but wife does all the cooking/planning. White British	***Gudrun*** (63), single woman, lives alone in a bungalow on a campsite, receives victim support and state benefits (Hartz IV), university educated, also completed apprenticeship in bakery/confectionary and is a trained equestrian instructor

UK	Germany
Melanie and Tom (late 20s). Both in full-time employment. White British ***Laurentio*** (late 30s). In full-time employment, living in a house-share. White British (Italian heritage)	***Stephan***, mid-40s, vice-president at a tele-comms company ***Greta*** (late 70s), grew up in Bonn, married twice, but divorced and now lives on her own. Recently moved to Remagen, lives in city centre. Previously worked as an estate agent
Michael (late 30s). Full-time employed, has a long-term partner but live in separate households. White British	***Michelle and Kirsten***, food bank customers

The canteen study:

Canteen guests	Female	Male
Persons undergoing rehabilitation, vocational training, apprenticeship	17 persons, aged 14–60	7 persons, aged 17–44
Instructors, consultants	3 persons, aged 38–40+	7 persons, aged 20–60
Judges	3 persons, aged 32–49	
Office employees dealing with administration, customers etc	19 persons, aged 21–59	6 persons, aged 21–65
Court reporter, judicial officer	2 persons, aged 57–64	
External guests	2 persons, a 60 year old owner of a small gallery and a 61 year old public employee	7 persons, a 21 years old unemployed man, a 56 year old male police officer, a 30 year old male athlete, a 51 year old unemployed man, a public employee in his late 50s, 2 retired men in their late 60s
Medical staff	16 persons, aged 28–63, including Carla and Petra	24 persons, aged 27–41, including Markus, Manfred, Klaus, Paul and Bernd
Janitor, craftsmen		7 persons, aged 25–63
Pre-school teacher	3 persons, aged 22–49	

The meal-box study:

Families who subscribe to The Seasons (organic) scheme	Families who subscribe to the Fairly Easy (non-organic) scheme
Bettina has a clerical background and works as a secretary at a clinic. Her husband is a biologist and has a PhD in toxicology. He works in a medical firm. They have two children aged 17 and 19 both living at home	***Marie*** is a physiotherapist, married to an orthopaedic surgeon. They have a girl aged 6 and a boy aged 8
Henrik has a degree in management and works as a manager at a laundry firm. His wife has a degree in HR and works as a secretary. He has two teenage boys from another marriage who live with Henrik part time	***Anja*** trained as a clerk and secretary and works as a medical secretary. Her husband is a trained blacksmith, technician and industrial electrician
Maiken trained as a primary school teacher and teaches maths and home economics. Her husband is an engineer and has a PhD. They have two boys aged 8 and 11	***Carsten*** is a trained laboratory technician. Her husband works in sales. They have two boys aged 3 and 7
Hanne is a structural engineer and works as a Department Manager. Her partner has degrees in mechanical engineering and financial management and works as a Department Manager. They have one boy aged 11 who has a form of Asperger syndrome	***Thilde*** has a bachelor's degree in pedagogy and works as head of department in a IT-company. Her husband has a master's degree in agricultural economics and works as a Functional Manager for the state. They have a boy aged 11 and a girl aged 9
Rikke is a veterinarian. His partner works as an auto electro-mechanic. They have two boys aged 6 and 10	***Helene*** is a doctor, working as a chief physician at an intensive care unit in a hospital. Her husband worked in the army where he took a Defence Officer Training programme. He now works as a major in the army. They have a boy aged 15 and a girl aged 13
Jytte runs a local boxing club with her husband who was trained as a mechanic. They have twins, a boy and a girl aged 15, and an adopted daughter aged 4	***Agnes*** has a bachelor in engineering and works as an operations manager. Her husband has a PhD in engineering and has his own business. They have three children: a girl aged 7 and two boys aged 12 and 13

Families who subscribe to The Seasons (organic) scheme	Families who subscribe to the Fairly Easy (non-organic) scheme
Gro has a Bachelor's degree in Business, Language and Culture. Her husband has a secondary education. Both parents work full time. They have two daughters aged 12 and 15—the oldest living at a boarding school at the time of the interview (*efterskole*)	

References

Afflerback, S., Carter, S., Anthony, A., & Grauerholtz, L. (2013). Infant feeding: Consumerism in the age of intensive mothering and risk society. *Journal of Consumer Culture, 13*(3), 1–19.

Ahlgren, M., Gustafsson, I.-B., & Hall, G. (2004). Attitudes and beliefs directed towards ready-meal consumption. *Food Service Technology, 4*, 159–169.

Akrich, M. (1992). The de-scription of technical objects. In W. Bijker & J. Law (Eds.), *Shaping technology/building society: Studies in sociotechnical change* (pp. 205–224). Cambridge, MA: MIT Press.

Akrich, M., & Latour, B. (1992). A summary of a convenient vocabulary for the semiotics of human and nonhuman assemblies. In W. Bijker & J. Law (Eds.), *Shaping technology building society: Studies in sociotechnical change* (pp. 259–264). Cambridge, MA: MIT Press.

Alkerwi, A., Crichton, G., & Hébert, J. (2015). Consumption of ready-made meals and increased risk of obesity. *British Journal of Nutrition, 113*(2), 270–277.

Appadurai, A. (1986). *The social life of things: Commodities in cultural perspective*. Cambridge: Cambridge University Press.

Apple, R. (1987). *Mothers and medicine: A social history of infant feeding, 1890–1950*. Madison: University of Wisconsin Press.

Apple, R. (1995). Constructing motherhood: Scientific motherhood in the nineteenth and twenieth centuries. *Social History of Medicine, 8*(2), 161–178.

P. Jackson et al., *Reframing Convenience Food*,
https://doi.org/10.1007/978-3-319-78151-8

Apple, R. (2006). *Perfect motherhood: Science and childrearing in America*. New Brunswick: Rutgers University Press.

Arnould, E. J. (1989). Toward a broadened theory of preference formation and the diffusion of innovations: Cases from Zinder Province, Niger Republic. *Journal of Consumer Research, 16*(2), 239–267.

Askegaard, S., Ordabayeva, N., Chandon, P., Cheung, T., Chytkova, Z., Cornil, Y., Corus, C., Edell, J. A., Mathras, D., Junghans, A. F., Kristensen, D. G., Mikkonen, I., Miller, E. G., Sayarh, N., & Werle, C. (2014). Moralities in food and health research. *Journal of Marketing Management, 30*(17–18), 1800–1832.

Atkins, P. J., Lummel, P., & Oddy, D. J. (Eds.). (2007). *Food and the city in Europe since 1800*. London: Routledge.

Atkinson, L. (2014). Green moms: The social construction of green mothering via environmental advertising appeals. *Consumption, Markets and Culture, 17*(6), 553–572.

Atkinson, R. (1993). Humour in philosophy. In K. Cameron (Ed.), *Humour and history* (pp. 10–20). Oxford: Intellect.

Autio, M., Collings, R., Wahlen, S., & Anttila, M. (2013). Consuming nostalgia? The appreciation of authenticity in local food production. *International Journal of Consumer Studies, 37*, 564–568.

Bava, C., Jaeger, S., & Park, J. (2008). Constraints upon food provisioning practices in 'busy' women's lives: Trade-offs which demand convenience. *Appetite, 50*, 486–498.

Belasco, W. (2007). *Appetite for change: How the counterculture took on the food industry*. Ithaca and London: Cornell University Press.

Belasco, W. (2008). *Food: The key concepts*. Oxford: Berg.

Bell, D., & Valentine, G. (1997). *Consuming Geographies: We are where we eat*. London: Routledge.

Bentley, A. (2014). *Inventing baby food: Taste, health, and the industrialization of the American diet*. Berkeley and Los Angeles: University of California Press.

Berker, T., Hartmann, M., Punie, Y., & Ward, K. (2006). *Domestication of media and technology*. Glasgow: Open University Press.

Beskæftigelsesministeriet. (2011). *Kvinder og mænd på arbejdsmarkedet 2011* [Women and men in the labour market 2011]. Copenhagen: Ministry of Employment.

Biltekoff, C. (2002). 'Strong men and women are not products of improper food': Domestic science and the history of eating and identity. *Journal for the Study of Food and Society, 6*(1), 60–69.

Blake, M., Mellor, J., & Crane, L. (2010). Buying local food: Shopping practices, place, and consumption networks in defining food as 'local'. *Annals of the Association of American Geographers, 100*(2), 409–426.
BMEL. (2017). *Deutschland wie es isst: Der BMEL-Ernährungsreport 2017*. Berlin: Bundesministerium für Ernährung und Landwirtschaft (Federal Ministry of Food and Agriculture).
Botonaki, A., & Mattas, K. (2010). Revealing the values behind convenience food consumption. *Appetite, 55*, 629–638.
Braudel, F. (1958). Histoires et sciences sociales: la longue durée. *Annales: Économies, Sociétés, Civilisations, 13*(4), 725–753.
Braun, J., & Beckie, M. A. (2014). Against the odds: The survival of traditional food knowledge in a rural Albertan community. *Canadian Food Studies, 1*, 54–71.
Brembeck, H. (2005). Home to McDonald's: Upholding the family dinner with the help of McDonald's. *Food, Culture and Society, 8*(2), 215–226.
Brembeck, H. (2007). *Hem till McDonald's*. Stockholm: Carlssons.
Brembeck, H. (2012). Cozy Friday: An analysis of family togetherness and ritual overconsumption. In B. Czarniawska & O. Löfgren (Eds.), *Managing overflow in affluent societies* (pp. 125–140). New York: Routledge.
Brembeck, H., & Fuentes, M. (2017). Convenient food for baby: A study of weaning as a social practice. *Food, Culture and Society, 20*(4), 569–586.
Brembeck, H., Karlsson, M., Ossiansson, E., Shanahan, H., Jonsson, L., Bergström, K., & Ölander, E. (2010). *Ju mer vi är tillsammans: fyrtiotalisterna och maten* [The more we are together: Baby-boomers and food]. Stockholm: Carlssons.
Brembeck, H., Torell, V. B., Falkström, M., & Johansson, B. (Eds.). (2001). *Det konsumerande barnet* [The consuming child]. Gothenburg: Etnologiska Föreningen i Västsverige.
Bremmer, J., & Roodenburg, H. (1997). *A cultural history of humour: From antiquity to the present day*. Cambridge: Polity Press.
Brown, A., & Lee, M. (2011). A descriptive study investigating the use and nature of baby-led weaning in a UK sample of mothers. *Maternal and Child Nutrition, 7*(1), 34–47.
Browne, A. L. (2016). Can people talk together about their practices? Focus groups, humour, and the sensitive dynamics of everyday life. *Area, 48*(2), 198–205.
Bruegel, M. (2002). How the French learned to eat canned food, 1809–1930s. In W. Belasco & P. Scranton (Eds.), *Food nation: Selling taste in consumer societies* (pp. 113–130). New York: Routledge.

Brunner, T. A., van der Horst, K., & Siegrist, M. (2010). Convenience food products: Drivers for consumption. *Appetite, 55*, 498–506.

Buckley, M., Cowan, C., & McCarthy, M. (2007). The convenience food market in Great Britain: Convenience food lifestyle (CFL) segments. *Appetite, 49*, 600–617.

Bugge, A. B., & Almås, R. (2006). Domestic dinner: Representations and practices of a proper meal among young suburban mothers. *Journal of Consumer Culture, 6*, 203–228.

Buhl, B. (2010). *Historien om danskernes mad i 15.000 år* [The history of the food of the Danes over 15,000 years]. Copenhagen: Dansk Landbrugsmuseum.

Burridge, J., & Barker, M. (2009). Food as a medium for emotional management of the family: Avoiding complaint and producing love. In P. Jackson (Ed.), *Changing families, changing food* (pp. 146–164). Basingstoke: Palgrave Macmillan.

Cabinet Office. (2008). *Food matters: Towards a strategy for the 21st century.* London: Cabinet Office Strategy Unit.

Caraher, M., Dixon, P., Lang, T., & Carr-Hill, R. (1999). The state of cooking in England: The relationship of cooking skills to food choice. *British Food Journal, 101*(8), 590–609.

Carrigan, M., & Szmigin, I. (2006). 'Mothers of invention': Maternal empowerment and convenience consumption. *European Journal of Marketing, 40*(9–10), 1122–1142.

Carrigan, M., Szmigin, I., & Leek, S. (2006). Managing routine food choices in UK families: The role of convenience consumption. *Appetite, 47*, 371–383.

Cheng, S.-L., Olsaen, W., Southerton, D., & Warde, A. (2007). The changing practice of eating: Evidence from UK time diaries, 1975 and 2000. *British Journal of Sociology, 58*, 39–61.

Cochoy, F. (2004). Is the modern consumer a Buridan's donkey? Product packaging and consumer choice. In K. M. Ekström & H. Brembeck (Eds.), *Elusive consumption* (pp. 45–64). Oxford: Berg.

Cochoy, F. (2005). A brief history of 'customers', or the gradual standardization of markets and organizations. *Sociologie du travail, 47*, e36–e56.

Cook, D. T. (1995). The mother as consumer. *The Sociological Quarterly, 36*(3), 505–522.

Coop. (2017). Sundere at leve nemt [Healthier to live easily]. Retrieved December 10, 2017, from https://om.coop.dk/sammen+om+bedre+mad/sundere+at+leve+nemt.aspx.

Coop-Analyse. (2016). Convenience vækster [Convenience grows], 12 July. Retrieved April 4, 2018, from https://coopanalyse.dk/analyse/convenience-v%C3%A6kster.

Costa, A., Dekker, M., Beuner, R., Rombauts, F., & Jongen, W. (2001). A consumer-oriented classification system for home replacement meals. *Food, Quality and Preference, 12*, 229–242.

Costa, A. I. D., Schoolmeester, D., Dekker, M., & Jongen, W. M. F. (2007). To cook or not to cook: A means-end study of motives for choice of meal solutions. *Food Quality and Preference, 18*, 77–88.

Counihan, C. (2004). *Around the Tuscan Table: Food, family and gender in twentieth-century Florence*. London: Routledge.

Coveney, J. (2000). *Food, morals and meaning: The pleasure and anxiety of eating*. London: Routledge.

Coveney, J., Begley, A., & Gallegos, D. (2012). 'Savoir fare': Are cooking skills a new morality? *Australian Journal of Adult Learning, 52*, 617–642.

Cowan, R. S. (1983). *More work for mother: The ironies of household technology from the open hearth to the microwave*. New York: Basic Books.

CPH Post. (2015). Danes buying more convenience food. Retrieved December 15, 2017, from http://cphpost.dk/news/danes-buying-more-convenience-food.html.

Cresswell, T. (2004). *Place: A short introduction*. Oxford: Blackwell.

Damberg, J. (2015). *Nu äter vi!: de moderna favoriträtternas okända historia* [Now we eat]. Stockholm: Ponto Pocket.

Daniels, S., & Glorieux, I. (2015). Convenience, food and family lives: A socio-typological study of household food expenditures in 21st-century Belgium. *Appetite, 94*, 54–61.

Daniels, S., Glorieux, I., Minnen, J., & van Tienoven, T. P. (2012). More than preparing a meal? Concerning the meanings of home cooking. *Appetite, 58*(3), 1050–1056.

de Boer, M., McCarthy, M., Cowan, C., & Ryan, I. (2004). The influence of lifestyle characteristics and beliefs about convenience food on the demand for convenience foods in the Irish market. *Food Quality and Preference, 15*, 155–156.

Deding, M., Lausten, M., & Andersen, A. (2006). *Børnefamiliernes balance mellem familie- og arbejdsliv* [The balance between family-life and work-life in families with young children]. Copenhagen: Social Forsknings Instituttet [Social Research Institute].

Defra. (2005). *The validity of food miles as an indicator of sustainable development: Final report*. London: Department for Environment, Food and Rural Affairs.

Defra. (2012). *Green food project: Conclusions*. London: Department for Environment, Food and Rural Affairs.

Defra. (2013). *Sustainable consumption report: Follow-up to the green food project*. London: Department for Environment, Food and Rural Affairs.

Delaney, M., & McCarthy, M. B. (2014). Saints, sinners and non-believers: The moral space of food. A qualitative exploration of beliefs and perspectives on healthy eating of Irish adults aged 50–70. *Appetite, 73*, 105–113.

DeVault, M. (1991). *Feeding the family: The social organisation of caring as gendered work*. Chicago: University of Chicago Press.

Deville, J., Guggenheim, M., & Hrdličková, Z. (2016). Same, same but different: Provoking relations, assembling the comparator. In J. Deville, M. Guggenheim, & Z. Hrdličková (Eds.), *Practising comparison: Logics, relations, collaborations* (pp. 99–129). Manchester: Mattering Press.

Dixon, J. M., Hinde, S. J., & Banwell, C. L. (2006). Obesity, convenience and 'phood'. *British Food Journal, 108*(8), 634–645.

Du Puis, E. M. (2015). *Dangerous digestion: The politics of American dietary advice*. Oakland, CA: University of California Press.

Ekstrom, M. P., & d'Orange Furst, E. (2001). The gendered division of cooking. In U. Kjaernes (Ed.), *Eating patterns: A day in the life of Nordic peoples* (pp. 213–234). Oslo: SIFO.

Eriksson, L. (2004). *Korv, mos och människor* [Sausage, mash and people]. Stockholm: Wahlström och Widstrand.

Euromonitor International. (2016a). Ready meals in Denmark. Retrieved December 15, 2017, from www.euromonitor.com/ready-meals-in-denmark/report.

Euromonitor International. (2016b). Ready meals in Sweden. Retrieved December 15, 2017, from www.euromonitor.com/ready-meals-in-sweden/report.

Evans, B., & Colls, R. (2009). Measuring fatness, governing bodies: The spatialities of the Body Mass Index (BMI) in anti-obesity politics. *Antipode, 41*, 1051–1083.

Evans, D. (2011). Blaming the consumer—Once again: The social and material contexts of everyday food waste practices in some English households. *Critical Public Health, 21*(4), 429–440.

Evans, D. (2012). Beyond the throwaway society: Ordinary domestic practice and a sociological approach to household food waste. *Sociology, 46*(1), 41–56.

Evans, D. (2014). *Food waste: Home consumption, material culture and everyday life*. London: Bloomsbury.

Evans, D., & Welsh, D. (2013). *Food waste transitions: Consumption, retail and collaboration towards a sustainable food system*. Report from the Sustainable Consumption Institute, University of Manchester.

Everts, J., & Jackson, P. (2009). Modernisation and the practices of contemporary food shopping. *Environment and Planning D: Society and Space, 27*, 917–935.

Fagerfjäll, R. (2012). *Semper: vällingen som blev bestående* [Semper: Porridge that became pemanent]. Stockholm: Centrum för Näringslivshistoria.

Feasey, R. (2008). *Masculinity and popular television*. Edinburgh: Edinburgh University Press.

Ferguson, P. P. (2006). *Accounting for taste: The triumph of French cuisine*. Chicago: University of Chicago Press.

Findus. (1962). *The Findus saga*. Switzerland: Findus International Ltd.

Flyjbjerg, B. (2001). *Making social science matter: Why social inquiry fails and how it can succeed again*. Cambridge: Cambridge University Press.

Foresight. (2007). *Tackling obesities: Future choices*. London: Government Office for Science.

Forty, A. (1986). *Objects of desire: Designs and society 1750–1980*. London: Thames and Hudson.

Foucault, M. (1995). *Dicipline and punish: The birth of the prison*. New York: Vintage Books.

Fox, R., Nicholson, P., & Heffernan, K. (2009). Pregnancy police? Maternal bodies, surveillance and food. In P. Jackson (Ed.), *Changing families, changing food* (pp. 57–74). Basingstoke: Palgrave Macmillan.

Fox, R., & Smith, G. (2011). Sinner ladies and the gospel of good taste: Geographies of food, class and care. *Health & Place, 17*(2), 403–412.

Franck, K. A. (Ed.). (2005). *Food and the city*. London: Wiley.

Freidberg, S. (2004). *French beans and food scares: Culture and commerce in an anxious age*. Oxford: Oxford University Press.

Freidberg, S. (2009). *Fresh: A perishable history*. Boston: Harvard University Press.

FSA. (2008). *Criteria for the use of the terms fresh, pure, natural etc. in food labelling*. London: Food Standards Agency. Retrieved April 4, 2018, from https://www.food.gov.uk/sites/default/files/multimedia/pdfs/markcritguidance.pdf.

FSA. (2016). *Our food future: Literature review*. London: Food Standards Agency.

Fuentes, C., & Fuentes, M. (2017). Making a market for alternatives: Marketing devices and the qualification of a vegan milk substitute. *Journal of Marketing Management, 33*(7–8), 529–555.

Fuentes, M., & Brembeck, H. (2017). Best for baby? Framing weaning practice and motherhood in web-mediated marketing. *Consumption, Markets and Culture, 20*(2), 153–175.

Geeroms, N., Verbeke, W., & van Kenhove, P. (2008). Consumers' health-related motive orientations and ready meal consumption behaviour. *Appetite, 51*, 704–712.

Gerber, S. (2015). *Küche, Kühlschrank, Kilowatt: Zur Geschichte des privaten Energiekonsums in Deutschland 1945–1990*. Bielefeld: Transcript.

GFS. (2016). *Food systems dialogue: A GFS food futures panel activity*. OPM Group for the Global Food Security Programme.

Giard, L. (1998). Doing cooking. In M. De Certeau, L. Giard, & P. Mayoll (Eds.), *The practice of everyday life, Vol. 2: Living and cooking* (pp. 149–247). Minneapolis: University of Minnesota Press.

Gluck, C., & Tsing, A. L. (Eds.). (2009). *Words in motion: Towards a global lexicon*. Durham and London: Duke University Press.

Gofton, L. (1995). Dollar rich and time poor? Some problems in interpreting changing food habits. *British Food Journal, 97*(10), 11–16.

Gong, Q., & Jackson, P. (2012). Consuming anxiety? Parenting practices in China after the infant formula scandal. *Food, Culture and Society, 15*(4), 557–578.

Gong, Q., & Jackson, P. (2013). Mediating science and nature: Representing and consuming infant formula advertising in China. *European Journal of Cultural Studies, 16*, 285–309.

Goodman, D., & Watts, M. (Eds.). (1997). *Globalizing food: Agrarian questions and global restructuring*. London and New York: Routledge.

Grinnell-Wright, V., Wilson, J., & Downing, P. (2013). *Review of evidence on consumer food-related behaviours that impact on sustainability* (Final Report EVO541). Best Foot Forward: The Sustainability Consultants.

Gritsai, O. (2001). Haute cuisine versus healthy nutrition: Cultural gradients in Europe and the geography of baby food. *GeoJournal, 53*, 71–80.

Grunert, K. G. (2003). Conveniencemad [Convenience food]. In L. Holm & S. T. Kristensen (Eds.), *Mad, Mennesker og Måltider [Food, people and meals]* (pp. 227–238). Copenhagen: Munksgaard (second edition 2012).

Guthman, J. (2011). *Weighing in: Obesity, food justice, and the limits of capitalism*. Berkeley and Los Angeles: University of California Press.

Hagberg, J.-E., & Kaijser, A. (1987). *Lotta—mellan diskmaskinen och framtiden: en bok om vardagslivets tekniska villkor* [Lotta: Between the dishwasher and the future: A book about everyday living conditions]. Linköping: Universitetet i Linköping.

Halkier, B. (1999). Consequences of the politicization of consumption: The example of environmentally friendly consumption practices. *Journal of Environmental Policy and Planning, 1*, 25–41.

Halkier, B. (2009). Suitable cooking? Performances and positionings in cooking practices among Danish women. *Food, Culture and Society, 12*, 357–377.

Halkier, B. (2010). *Consumption challenged: Food in medialised everyday lives*. Farnham: Ashgate.

Halkier, B. (2013). Easy eating? Negotiating convenience food in media food practices. In L. Hansson, U. Holmberg, & H. Brembeck (Eds.), *Making sense of consumption* (pp. 119–136). Gothenburg: Gothenburg University Press.

Halkier, B. (2016). Civic engagement by invitation? Citizen negotiations about public media framings of everyday life responsibilities for societal problems. In H. K. Nielsen (Ed.), *The democratic public sphere: Current challenges and prospects* (pp. 289–317). Aarhus: Aarhus University Press.

Halkier, B. (2017a). Normalising convenience food? *Food, Culture and Society, 20*(1), 133–151.

Halkier, B. (2017b). Mundane science use in a practice theoretical perspective: Different understandings of the relations between citizen-consumers and public communication initiatives build on scientific claims. *Public Understanding of Science, 26*, 40–54.

Halkier, B., & Jensen, I. (2011). Doing 'healthier' food in everyday life? A qualitative study of how Pakistani Danes handle nutritional communication. *Critical Public Health, 21*(4), 471–483.

Halkier, B., Katz-Gerro, T., & Martens, L. (2011). Applying practice theory to the study of consumption: Theoretical and methodological considerations. *Journal of Consumer Culture, 11*(1), 3–13.

Hammersley, M., & Atkinson, P. (1995). *Ethnograpy: Principles in practice*. London: Routledge.

Hand, M., & Shove, E. (2007). Condensing practices: Ways of living with a freezer. *Journal of Consumer Culture, 7*(1), 79–104.

Hand, M., Shove, E., & Southerton, D. E. (2005). Explaining showering: A discussion of the material, conventional, and temporal dimensions of practice. *Sociological Research Online, 10*(2), 1–13.

Hareven, T. K. (1982). *Family time and industrial time: The relationship between the family and work in a New England industrial community*. Cambridge: Cambridge University Press.

Harvey, D. (1990). Between space and time: Reflections on the geographical imagination. *Annals of the Association of American Geographers, 80*(3), 418–434.

Hawkins, G. (2011). Packaging water: Plastic bottles as market and public devices. *Economy and Society, 40*(4), 534–552.

Hawkins, G. (2012). The performativity of food packaging: Market devices, waste crisis and recycling. *The Sociological Review, 60*(2), 66–83.

Hecht, G. (1998). *The radiance of France: Nuclear power and national identity after World War II*. Cambridge, MA: MIT Press.

Hecht, G. (2009). *Entangled geographies: Empire and technopolitics in the Global Cold War*. Cambridge, MA: MIT Press.

Heinzelmann, U. (2014). *Beyond bratwurst: A history of food in Germany*. London: Reaktion Books.

Heldke, L. (2003). *Exotic appetites: Ruminations of a food adventurer*. London: Routledge.

Henderson, J., Ward, P., Coveney, J., & Taylor, A. (2009). 'Heath is the number one thing we go for': Healthism, citizenship and food choice. In S. Lockie, D. Bissell, A. Greig, M. Hynes, D. Marsh, L. Saha, J. Sikora, & D. Woodman (Eds.), *The future of sociology*. Canberra, ACT: The Australian Sociological Association.

Hertz, F. D., & Halkier, B. (2017). Meal box schemes as a convenient way to avoid convenience food? Uses and understandings of meal box schemes among Danish consumers. *Appetite, 114*, 232–239.

Heßler, M. (2001). 'Elektrische Helfer' für Hausfrau, Volk und Vaterland: Ein technisches Konsumgut während des Nationalsozialismus. *Technikgeschichte, 68*(3), 203–229.

Hirschfelder, G. (2001). *Europäische Esskultur: Geschichte der Ernährung von der Steinhzeit bis heute*. Frankfurt am Main: Campus Verlag.

Holden, T. J. M. (2005). The overcooked and the underdone: Masculinities in Japanese food programming. *Food and Foodways, 13*, 39–65.

Hollows, J. (2003a). Oliver's twist: Leisure, labour and domestic masculinity in The Naked Chef. *International Journal of Cultural Studies, 6*, 229–248.

Hollows, J. (2003b). Feeling like a domestic goddess: Postfeminism and cooking. *European Journal of Cultural Studies, 6*, 179–202.

Hollows, J., & Jones, S. (2010). 'At least he's doing something': Moral entrepreneurship and individual responsibility in Jamie's Ministry of Food. *European Journal of Cultural Studies, 13*(3), 307–322.

Holm, L. (2003). Blaming the consumer: On the free choice of consumers and the decline in food quality in Denmark. *Critical Public Health, 13*, 139–154.

Holm, L., Ekstrøm, M. P., Hach, S., & Lund, T. B. (2016). Who is cooking dinner? *Food, Culture and Society, 18*, 589–610.

Holtzman, J. (2006). Food and memory. *Annual Review of Anthropology, 35*, 361–378.

van der Horst, K., Brunner, T. A., & Siegrist, M. (2010). Ready-meal consumption: Associations with weight status and cooking skills. *Public Health Nutrition, 14*, 239–245.

Horton, P., Banwart, S. A., Brockington, D., Brown, G. W., Bruce, R., Cameron, D., Holdsworth, M., Koh, S. C. L., Ton, J., & Jackson, P. (2017). An agenda for integrated system-wide interdisciplinary agri-food research. *Food Security, 9*(2), 195–210.

Howard, S., Adams, J., & White, M. (2012). Nutritional content of supermarket readymeals and recipes by television chefs in the United Kingdom: Cross sectionalstudy. *British Medical Journal, 345*, e7607.

Hunt, A. (2003). Risk and moralization in everyday life. In R. Ericson & A. Doyle (Eds.), *Risk and morality* (pp. 165–192). Toronto: University of Toronto Press.

Inness, S. A. (2001). *Dinner roles: American women and culinary culture*. Iowa: University of Iowa Press.

Jabs, J., & Devine, C. M. (2006). Time scarcity and food choices: An overview. *Appetite, 47*, 196–204.

Jackson, P. (Ed.). (2009). *Changing families, changing food*. Basingstoke: Palgrave Macmillan.

Jackson, P. (2015). *Anxious appetites: Food and consumer culture*. London: Bloomsbury.

Jackson, P. (2016). Go Home Jamie: Reframing consumer choice. *Social and Cultural Geography, 17*(6), 1–5.

Jackson, P. (2018). Familial fictions: Families and food, convenience and care. *European Journal of Marketing*. https://doi.org/10.1108/EJM-11-2017-0882.

Jackson, P., & Everts, J. (2010). Anxiety as social practice. *Environment and Planning A, 42*(11), 2791–2806.

Jackson, P., & Meah, A. (2018). Reassessing vulnerability to foodborne illness: Pathways and practices. *Critical Public Health, 28*(1), 81–93.

Jackson, P., Olive, S., & Smith, G. (2009). Myths of the family meal: Re-reading Edwardian life histories. In P. Jackson (Ed.), *Changing families, changing food* (pp. 131–145). Basingstoke: Palgrave Macmillan.

Jackson, P., Perez del Aguila, R., Clarke, I., Hallsworth, A., de Kervenoael, R., & Kirkup, M. (2006). Retail restructuring and consumer choice 2: Understanding consumer choice at the household level. *Environment and Planning A, 38*, 47–67.

Jackson, P., & Viehoff, V. (2016). Reframing convenience food. *Appetite, 98*, 1–11.

Jaeger, S., & Meiselman, H. (2004). Perceptions of meal convenience: The case of at-home evening meals. *Appetite, 42*, 317–325.

Jain, A., Sherman, S. N., Chamberlin, L. A., & Carter, Y. (2001). Why don't low-income mothers worry about their preschoolers being overweight? *Pediatrics, 107*(5), 1138–1146.

Janhonen, K. (2017). The roles of humour and laughter in youth focus groups on school food. *Journal of Youth Studies, 20*(9), 1127–1142.

Janhonen, K. What's so funny about school food? Functions of laughter in young people's group interaction. *Young: Nordic Journal of Youth Research*, in review.

Jelsma, J. (2003). Innovation for sustainability: Involving users, politics and technology. *Innovation: The European Journal of Social Science Research, 16*(2), 103–116.

Jelsma, J. (2006). Designing 'moralized' products: Theory and practice. In P. P. Verbeek & A. Slob (Eds.), *User behaviour and technology development: Shaping sustainable relationships between consumers and technologies* (pp. 221–231). Dordrecht: Springer.

Kahma, N., Mäkelä, J., Niva, M., Ganskau, E., & Minina, V. (2016). Convenience food consumption in the Nordic countries and St. Petersburg area. *International Journal of Consumer Studies, 40*(4), 492–500.

Keane, A. (1997). Too hard to swallow? The palatability of health eating advice. In P. Caplan (Ed.), *Food, health and identity* (pp. 172–192). London: Routledge.

Keenan, J., & Stapleton, H. (2010). Bonny babies? Motherhood and nurturing in the age of obesity. *Health, Risk & Society, 12*(4), 369–383.

Keenan, J., & Stapleton, H. (2014). 'It won't do her any harm' they said, 'or they wouldn't put it on the market': Infant weaning, markets and mothers' narratives of trust. In S. O'Donohoe, M. Hogg, P. Maclaran, L. Martens, & L. Stevens (Eds.), *Motherhood, markets and consumption: The making of mothers in contemporary Western cultures* (pp. 71–89). London: Routledge.

Korsmeyer, C. (2011). *Savoring disgust: The foul and the fair in aesthetics*. Oxford: Oxford University Press.

Krause, M. (2016). Comparative research: Beyond linear-causal explanation. In J. Deville, M. Guggenheim, & Z. Hrdličková (Eds.), *Practising comparison: Logics, relations, collaborations* (pp. 45–67). Manchester: Mattering Press.

Kreyenfeld, M., & Geisler, E. (2006). Müttererwerbstätigkeit in Ost- und Westdeutschland. *Zeitschrift für Familienforschung, 18*(3), 333–360.

Lamont, M., & Thévenot, L. (2000). *Rethinking comparative cultural sociology: Repertoires of evaluation in France and the United States*. Cambridge: Cambridge University Press.

Lane, G. S., & Watson, G. L. (1975). A Canadian replication of Mason Haire's 'shopping list' study. *Journal of the Academy of Marketing Science, 3*(1), 48–59.

Lang, T., & Caraher, M. (2001). Is there a culinary skills transition? Data and debate from the UK about changes in cooking culture. *Journal of the HEIA, 8*, 2–14.

Lavelle, F., McGowan, L., Spence, M., Caraher, M., Raats, M., Hollywood, L., McDowell, D., McCloat, A., Mooney, E., & Dean, M. (2016). Barriers and

facilitators to cooking from 'scratch' using basic or raw ingredients: A qualitative interview study. *Appetite, 107*, 383–391.
Lavin, C. (2013). *Eating anxiety: The perils of food politics*. Minneapolis and London: University of Minnesota Press.
Lavis, A., Abbots, E.-J., & Attala, L. (2015). Introduction: Reflecting on the embodied intersection of eating and caring. In E.-J. Abbots, A. Lavis, & L. Attala (Eds.), *Careful eating: Bodies, food and care* (pp. 16–28). New York: Routledge.
Leer, J., & Povlsen, K. K. (2016). *Food and media: Practices, distinctions and heterotopias*. London: Routledge.
Lefebvre, H. (1991). *The production of space*. Oxford: Blackwell.
Lehrer, A. (1969). Semantic cuisine. *Journal of Linguistics, 5*(1), 39–55.
Levenstein, H. (1988). *Revolution at the table: The transformation of the American diet*. New York: Oxford University Press.
Lien, M. E. (1997). *Marketing and modernity: An ethnography of marketing practice*. Oxford: Berg.
Lin, X., Zhu, M., & Nettleton, S. (2017). Enduring 'care' and the shifting cultural meanings of convenience food. *Discover Society*, 6 December. Retrieved April 4, 2018, from https://discoversociety.org/2017/12/06/enduring-care-and-the-shifting-cultural-meanings-of-convenience-food/.
Livsmedelsverket. (2011). *Råd om mat för barn 0-5 år* [Advice about food for children 0–5 years]. Rapport 22-2011.
Livsmedelsverket. (2015). *Råd om mat för barn 0-5 år: handledning för barnhälsovården*. Retrieved December 17, 2017, from https://www.livsmedelsverket.se/globalassets/matvanor-halsa-miljo/kostrad-matvanor/spadbarn/bra-mat-for-barn-0-5-ar---handledning-for-barnhalsovarden.pdf.
Luks, T. (2009). Die Massengesellschaft auf dem Weg in die Kantine: Fabrikmahlzeit, Selbstbedienung und 'Ordnungsdenken' bei der Daimler-Benz AG 1948-53. *Historische Anthropologie, 17*(1), 38–55.
Lupton, D. (1994). Food, memory and meaning: The symbolic and social nature of food events. *The Sociological Review, 42*(4), 664–685.
Lyon, P., Colquhoun, A., & Alexander, E. (2003). Deskilling the domestic kitchen: National tragedy or the making of a modern myth? *Food Service Technology, 3*, 167–175.
Macdiarmid, J. I., Douglas, F., & Campbell, J. (2016). Eating like there's no tomorrow: Public awareness of the environmental impact of food and reluctance to eat less meat as part of a sustainable diet. *Appetite, 96*, 487–493.
Mahon, D., Cowan, C., & McCarthy, M. (2006). The role of attitudes, subjective norm, perceived control and habit in the consumption of ready meals

and takeaways in Great Britain. *Food Quality and Preference, 17*(6), 474–481.
Mäkelä, J., Lillebø, K., & Lammi, M. (Eds.). (2011). *Nordic young health possibilities and barriers for new, healthy concepts in the fast food sector*. Oslo: Nordic Innovation Report.
Mandemakers, J. J., & Roeters, A. (2015). Fast or slow food? Explaining trends in food-related time in the Netherlands, 1975–2005. *Acta Sociologica, 58*, 121–137.
Marshall, D., & Bell, R. (2003). Meal construction: Exploring the relationship between eating occasion and location. *Food Quality and Preference, 14*(1), 53–64.
Marshall, J., Godfrey, M., & Renfrew, M. J. (2007). Being a 'good mother': Managing breastfeeding and merging identities. *Social Science and Medicine, 65*, 2147–2159.
Massey, D. (2004). Geographies of responsibility. *Geografiska Annaler B, 86*(1), 5–18.
Massey, D. (2005). *For space*. London: Sage.
Max Rubner-Institut. (2008). *Nationale Verzehrsstudie II*. Karlsruhe: Max Ribner-Institut.
May, C., & Finch, T. (2009). Implementing, embedding, and integrating practices: An outline of normalization process theory. *Sociology, 43*(3), 535–554.
Meah, A. (2014). Still blaming the consumer? Geographies of responsibility in domestic food safety practices. *Critical Public Health, 24*, 88–103.
Meah, A. (2016). Materializing memory, mood, and agency: The emotional geographies of the modern kitchen. *Gastronomica, 16*(2), 55–68.
Meah, A. (2017). Fathers, food practices and the circuits of intimacy in families in Northern England. *Gender, Place and Culture, 24*(8), 1145–1164.
Meah, A., & Jackson, P. (2013). Crowded kitchens: The 'democratisation' of domesticity? *Gender, Place and Culture, 20*(5), 578–596.
Meah, A., & Jackson, P. (2016). Re-imagining the kitchen as a site of memory. *Social and Cultural Geography, 17*(4), 511–532.
Meah, A., & Jackson, P. (2017). Convenience as care: Culinary antinomies in practice. *Environment and Planning A, 49*(9), 2065–2081.
Meah, A., & Watson, M. (2011). Saints and slackers: Challenging discourses about the decline of domestic cooking. *Sociological Research Online, 16*(2), 1–13.
Meah, A., & Watson, M. (2013). Cooking up consumer anxieties about 'provenance' and 'ethics': Why it sometimes matters where foods come from in domestic provisioning. *Food, Culture and Society, 16*(3), 495–512.
Mennell, S. (1987). On the civilizing of appetite. *Theory, Culture and Society, 4*(2), 373–403.

Merkel, I. (1998). Consumer culture in the GDR, or how the struggle for anti-modernity was lost in the battleground of consumer culture. In S. Strasser, C. McGovern, & M. Judt (Eds.), *Getting and spending: European and American consumer societies in the twentieth century* (pp. 2633–2280). Cambridge: Cambridge University Press.

Miller, D. (1998). *A theory of shopping*. Cambridge: Polity Press.

Milne, R. J. (2013). Arbiters of waste: Date labels, the consumer and knowing good safe food. In D. Evans, H. Campbell, & A. Murcott (Eds.), *Waste matters: New perspectives on food and society* (pp. 84–101). Oxford: John Wiley & Sons (Sociological Review Monographs).

MINTEL. (2010). *Ready to eat meals: A revolution*. London: Mintel Group.

MINTEL. (2013). *Prepared meals*. London: Mintel Group.

Mitchell, J. C. (1983). Case and situation analysis. *Sociological Review, 31*, 187–211.

Moisio, R., Arnould, E. J., & Price, L. L. (2004). Between mothers and markets: Constructing family identity through home-made food. *Journal of Consumer Culture, 4*(3), 361–384.

Montelius, E., & Nygren, K. G. (2014). 'Doing' risk, 'doing' difference: Towards an understanding of the intersections of risk, morality and taste. *Health, Risk and Society, 16*(5), 431–443.

Moore, A. P., Milligan, P., Rivas, C., & Goff, L. M. (2012). Sources of weaning advice, comparisons between formal and informal advice, and associations with weaning timing in a survey of UK first-time mothers. *Public Health Nutrition, 15*, 1661–1669.

Murcott, A. (1983). 'It's a pleasure to cook for him': Food, mealtimes and gender in some South Wales households. In E. Gamarnikow, D. Morgan, J. Purvis, & D. Taylorson (Eds.), *The public and the private* (pp. 78–90). London: Heinemann.

Murcott, A. (1995). Raw, cooked and proper meals at home. In D. Marshall (Ed.), *Food, choice and the consumer* (pp. 219–234). Glasgow: Blackie and Academic.

Murcott, A. (1997). Family meals—A thing of the past? In P. Caplan (Ed.), *Food, health and identity* (pp. 32–49). London: Routledge.

Närvänen, E., Saarijärvi, H., & Simanainen, O. (2013). Understanding consumers' online conversation practices in the context of convenience food. *International Journal of Consumer Studies, 37*, 569–576.

Nielsen. (2006). *Consumers and ready-to-eat meals: A global A C Nielsen report*. Oxford: A C Nielsen.

Nielsen, J. K. (2017). Dansk detailhandels udvikling efter 1960 [The development of the Danish retail sector after 1960] in *Den Store Danske Encyklopædi* [The big Danish encyclopaedia]. Retrieved April 4, 2018, from http://denstoredanske.dk/.

Nordic Food Survey. (2015). *Nordic food survey 2015: Consumer trends*. London: Ernst & Young Global Ltd.

Oakley, A. (1985). *The sociology of housework*. Oxford: Blackwell.

O'Brien, M. C., McConnon, A., Hollywood, L. E., Cuskelly, G. J., Barnett, J., Raats, M., & Dean, M. (2015). Let's talk about health: Shoppers' discourse regarding health while food shopping. *Public Health Nutrition, 18*(6), 1001–1010.

Oldenziel, R., & Zachmann, K. (2009). Kitchens as technology and politics: An introduction. In R. Oldenziel & K. Zachmann (Eds.), *Cold war kitchen: Americanization, technology, and European users* (pp. 1–29). Cambridge, MA: MIT Press.

Oliver, J., Loftus, D., & Terry, C. (2009). *Jamie's food revolution*. New York: Hyperion.

Olsen, N. V., Sijtsema, S. J., & Hall, G. (2010). Predicting consumers' intention to consume ready-to-eat meals: The role of moral attitude. *Appetite, 55*, 534–539.

Olsen, S. O., Prebensen, N., & Larsen, T. A. (2009). Including ambivalence as a basis for benefit segmentation: A study of convenience food in Norway. *European Journal of Marketing, 43*(5–6), 762–783.

Parkin, K. (2006). *Food is love: Advertising and gender roles in modern America*. Philadelphia: University of Pennsylvania Press.

Pfau, C., & Saba, A. (2009). Preparing meals in later life. In M. Raats, L. de Groot, & W. van Staveren (Eds.), *Food for the ageing population* (pp. 560–579). Cambridge: Woodhead Publishing Ltd.

Piper, N. (2013). Audiencing Jamie Oliver: Embarrassment, voyeurism and reflexive positioning. *Geoforum, 45*, 346–355.

Plessz, M., & Gojard, S. (2015). Fresh is best? Social position, cooking and vegetable consumption in France. *Sociology, 49*, 172–190.

Pollan, M. (2009). *Food rules: An eater's manual*. London: Penguin.

Prim, M. K., Gustafsson, I., & Hall, G. (2007). The appropriateness of ready meals for dinner. *Journal of Foodservice, 18*, 238–250.

Probyn, E. (2000). *Carnal appetites: Food sex identities*. London: Routledge.

Radio Sweden. (2013, January 12). Arsenic and lead found in Swedish baby food.

Ragin, C. (1987). *The comparative method: Moving beyond qualitative and quantitative strategies*. Berkeley: University of California Press.

Reckwitz, A. (2002). Toward a theory of social practices: A development in culturalist theorizing. *European Journal of Social Theory, 5*, 243–263.

Reed, Z., McIlveen, H., & Strugnell, C. (2000). The retailing environment in Ireland and its effect on the chilled ready meal market. *Journal of Consumer Studies and Home Economics, 24*, 234–241.

Reed, Z., McIlveen-Farley, H., & Strugnell, C. (2003). Factors affecting consumer acceptance of chilled ready meals on the island of Ireland. *International Journal of Consumer Studies, 27*, 2–10.

Reid, S. E. (2002). Cold war in the kitchen: Gender and the de-Stalinization of consumer taste in the Soviet Union under Khrushchev. *Slavic Review, 61*(2), 211–252.

Rich, E. (2011). 'I see her being obesed!': Public pedagogy, reality media and the obesity crisis. *Health, 15*(1), 3–21.

Ritzer, G. (1993). *The McDonaldization of society*. Thousand Oaks, CA: Pine Forge Press.

Rosa, H. (2013). *Social acceleration: A new theory of modernity*. New York: Columbia University Press.

Ross, K. (1995). *Fast cars, clean bodies: Decolonization and the reordering of French culture*. Cambridge, MA: The MIT Press.

Rousseau, S. (2012). *Food media*. London: Bloomsbury.

Rozin, P. (1999). The process of moralization. *Psychological Science, 19*(3), 218–221.

Samvirke. (2014). Hvad blev der af lørdagskyllingen? [What happened to the Saturday chicken?]. Retrieved December 15, 2017, from https://samvirke.dk/artikler/hvad-blev-der-af-lordagskyllingen.

Sassatelli, R. (2004). The political morality of food: Discourses, contestation and alternative consumption. In M. Harvey, A. McMeeken, & A. Warde (Eds.), *Qualities of food* (pp. 176–191). Manchester: Manchester University Press.

Sayer, A. (1992). *Method in social science: A realist approach* (2nd ed.). London: Routledge.

Schatzki, T. R. (1996). *Social practices: A Wittgensteinian approach to human activity and the social*. Cambridge: Cambridge University Press.

Schatzki, T. R. (2002). *The site of the social: A philosophical account of the constitution of social life and change*. Philadelphia: Pennsylvania University Press.

Schatzki, T. R. (2005). Peripheral vision: The sites of organizations. *Organization Studies, 26*(3), 465–484.

Schatzki, T. R. (2009). Timespace and the organization of social life. In E. Shove, F. Trentmann, & R. Wilk (Eds.), *Time, consumption and everyday life: Practice, materiality and culture* (pp. 35–48). Oxford: Berg.

Scholderer, J., & Grunert, K. (2005). Consumers, food and convenience: The long way from resource constraints to actual consumption patterns. *Journal of Economic Psychology, 26*, 105–128.
Scholliers, P. (2015). Convenience foods: What, why, and when. *Appetite, 94*, 2–6.
Schulze, G. (1997). From situations to subjects: Moral discourse in transition. In P. Sulkunen, J. Holmwood, H. Radner, & G. Schulze (Eds.), *Constructing the new consumer society* (pp. 38–57). Basingstoke: Macmillan.
Semper. (2016). Semper presenterar trendspaning om barnmat (press release). Retrieved December 20, 2017, from http://www.mynewsdesk.com/se/semper/pressreleases/semper-presenterar-trendspaning-om-barnmat-2016-1380569.
Sennett, R. (2008). *The craftsman*. New Haven, CT: Yale University Press.
Shapiro, L. (2004). *Something from the oven: Reinventing dinner in 1950s America*. New York: Penguin.
Shapiro, L. (2009). *Perfection salad: Women and cooking at the turn of the century*. New York: Farrar, Straus and Giroux.
Shaw, J. (2010). Philosophy of humour. *Philosophy Compass, 5*(2), 112–126.
Shephard, S. (2000). *Pickled, potted and canned: The story of food preserving*. London: Headline.
Short, F. (2006). *Kitchen secrets: The meaning of cooking in everyday life*. Oxford: Berg.
Shove, E. (2003). *Comfort, cleanliness and convenience*. Oxford: Berg.
Shove, E. (2010). Beyond the ABC: Climate change policy and theories of social change. *Environment and Planning A, 42*(6), 1273–1285.
Shove, E., & Pantzar, M. (2005). Consumers, producers and practices. Understanding the invention and reinvention of Nordic walking. *Journal of Consumer Culture, 5*(1), 43–64.
Shove, E., Pantzar, M., & Watson, M. (2012). *The dynamics of social practice: Everyday life and how it changes*. London: Sage.
Shove, E., & Southerton, D. (2000). Defrosting the freezer: From novelty to convenience. A narrative of normalization. *Journal of Material Culture, 5*(3), 301–319.
Shove, E., Trentmann, F., & Wilk, R. (Eds.). (2009). *Time, consumption and everyday life: Practice, materiality and culture*. Oxford: Berg.
Silva, E. (2000). The cook, the cooker and the gendering of the kitchen. *The Sociological Review, 48*(4), 612–628.
Silverman, D. (2006). *Interpreting qualitative data*. London: Sage.
Simmons, D., & Chapman, G. E. (2012). The significance of home cooking within families. *British Food Journal, 114*, 1184–1195.
Skibsted, L. H. (2017). Færdigretter [Ready meals], *Den Store Danske Encyklopædi* [The Great Danish Encyclopedia]. Gyldendal.

Slocum, R., Shannon, J., Cadieux, K. V., & Beckman, M. (2011). 'Properly, with love, from scratch': Jamie Oliver's food revolution. *Radical History Review, 110*, 178–191.

Smith, A. F. (Ed.). (2004). *Oxford Encyclopedia of Food and Drink in America* (Vol. 1). New York: Oxford University Press.

Southerton, D. (2003). Squeezing time: Allocating practices, coordinating networks and scheduling society. *Time & Society, 12*(1), 5–25.

Southerton, D. (2012). Habits, routines and temporalities of consumption: From individual behaviours to the reproduction of everyday practices. *Time & Society, 22*(3), 335–355.

Southerton, D., & Tomlinson, M. (2005). 'Pressed for time': The differential impacts of a 'time squeeze'. *The Sociological Review, 53*(2), 215–239.

Southerton, D., Diaz-Mendez, C., & Warde, A. (2011). Behavioural change and the temporal ordering of eating practices: A UK-Spain comparison. *International Journal of Sociology of Agriculture and Food, 19*(1), 19–36.

Statista. (2017). Kennzahlen der Verpflegung/Gastronomie in Betrieben in Deutschland. Retrieved December 4, 2017, from https://de.statista.com/statistik/daten/studie/680162/umfrage/kennzahlen-der-versorgung-in-unternehmen-in-deutschland/.

Strasser, S. (1982). *Never done: A history of American housework*. London: Macmillan.

Stuart, T. (2009). *Waste: Uncovering the global food scandal*. New York: WW Norton & Company.

Sutton, D. (2001). *Remembrance of repasts: An anthropology of food and memory*. Oxford: Berg.

Swenson, R. (2009). Domestic divo: Televised treatments of masculinity, femininity and food. *Critical Studies in Media Communication, 26*, 36–53.

Szabo, M. (2011). The challenges of 're-engaging with food': Connecting employment, household patterns and gender relations to convenience food consumption in North America. *Food, Culture and Society, 14*, 547–566.

Tanner, J. (1999). *Fabrikmahlzeit: Ernährungswissenschaft, Industriearbeit und Volksernährung in der Schweiz 1890–1950*. Zürich: Chronos Verlag.

Thelen, T. (2006). Lunch in an East German enterprise: Differences in eating habits as symbols of collective identities. *Zeitschrift für Ethnologie, 131*, 51–70.

Thompson, E. P. (1967). Time, work-discipline, and industrial capitalism. *Past & Present, 38*, 56–97.

Thoms, U. (2003). Industrial canteens in Germany, 1850–1950. In M. Jacobs & P. Scholliers (Eds.), *Eating out in Europe: Picnics, gourmet dining, and snacks since the late eighteenth century* (pp. 351–372). Oxford: Berg.

Thoms, U. (2009). Physical reproduction, eating culture and communication at the workplace: The case of industrial canteens in Germany 1850–1950. *Food & History, 7*(2), 119–153.

Thorndahl, J. (2001). *Fra saltkar og spisekammer til frostboks og køleskab—ny husholdningsteknologi indføres i Danmark* [From saltcellar and pantry to freezer and fridge—The introduction of new household technology in Denmark]. Elmuseet Årsskrift [The annual paper of the museum of electricity], 39–48.

Tilman, D., & Clark, M. (2014). Global diets link environmental sustainability and human health. *Nature, 515*, 518–522.

Trädgårdh Tornhill, C. (2012). *Barnmat: råd och recept.* Stockholm: Gothia Fortbildning.

Trentmann, F. (2005). Knowing consumers—Histories, identities, practices: An introduction. In F. Trentmann (Ed.), *The making of the consumer: Knowledge, power and identity in the modern world* (pp. 1–27). Oxford: Berg.

Trubek, A. B. (2000). *Haute cuisine: How the French invented the culinary profession.* Philadelphia: University of Pennsylvania Press.

Truninger, M. (2011). Cooking with Bimby in a moment of recruitment: Exploring conventions and practice perspectives. *Journal of Consumer Culture, 11*, 37–59.

Tuan, Y.-F. (1977). *Space and place: The perspective of experience.* Minneapolis: University of Minnesota Press.

Turner, K. L. (2006). Buying, not cooking. *Food, Culture and Society, 9*(1), 13–39.

Tyler, I. (2008). 'Chav mum chav scum': Class disgust in contemporary Britain. *Feminist Media Studies, 8*(1), 17–34.

Uhl, K. (2012). Schafft Lebensraum in der Fabrik: Betriebliche Kantinen und Speiseräume im deutschen Rationalisierungsdiskurs 1880–1945. In L. Bluma & K. Uhl (Eds.), *Kontrollierte Arbeit—disziplinierte Körper? Zur Sozial- und Kulturgeschichte der Industriearbeit im 19. und 20. Jahrhundert* (pp. 361–395). Bielefeld: Transcript (Histoire, 27).

Waddington, P. A. J. (1999). Police (canteen) sub-culture: An appreciation. *British Journal of Criminology, 39*(2), 287–309.

Wagner, K. (2013). 'Looks great, feels amazing': The tactile dimension of packaging. In L. Hansson, U. Holmberg, & H. Brembeck (Eds.), *Making sense of consumption* (pp. 139–153). Gothenburg: Centre for Consumer Science.

Wahlen, S., van der Horst, H., & Pothoff, R. (2016). How convenient!? Adolescents' vistas on food competences in a convenience context. *British Food Journal, 118*, 2828–2838.

Warde, A. (1997). *Consumption, food and taste: Culinary antinomies and commodity culture*. London: Sage.

Warde, A. (1999). Convenience food: Space and timing. *British Food Journal, 101*(7), 518–527.

Warde, A. (2005). Consumption and theories of practice. *Journal of Consumer Culture, 5*(2), 131–153.

Warde, A. (Ed.). (2010). *Consumption*. London: Sage.

Warde, A. (2014). After taste: Culture, consumption and theories of practice. *Journal of Consumer Culture, 14*(3), 279–303.

Warde, A. (2016). *The practice of eating*. Cambridge: Polity Press.

Warde, A., Cheng, S., Olsen, W., & Southerton, D. (2007). Changes in the practice of eating: A comparative analysis of time-use. *Acta Sociologica, 50*, 363–385.

Warde, A., & Martens, L. (2000). *Eating out: Social differentiation, consumption and pleasure*. Cambridge: Cambridge University Press.

Warde, A., & Yates, L. (2017). Understanding eating events: Snacks and meal patterns in Great Britain. *Food, Culture and Society, 20*(1), 15–36.

Watson, C. (2015). A sociologist walks into a bar (and other academic challenges): Towards a methodology of humour. *Sociology, 49*(3), 407–421.

Watson, M. (2013). Sustainability. In P. Jackson & The CONANX Group (Eds.), *Food words: Essays in culinary culture* (pp. 209–211). London: Bloomsbury.

Watson, M., & Meah, A. (2013). Food, waste and safety: Negotiating conflicting social anxieties into the practices of domestic provisioning. *The Sociological Review, 60*(S2), 102–120.

Which? (2015). *Food system challenges: Public dialogue on food system challenges and possible solutions.* TNS-BMRB for Which? and the Government Office for Science.

Wills, W. J., Meah, A., Dickinson, A. M., & Short, F. (2015). 'I don't think I ever had food poisoning': A practice-based approach to understanding foodborne disease that originates in the home. *Appetite, 85*, 118–125.

Wolfenstein, M. (1951). The emergence of fun morality. *Journal of Social Issues, 7*(4), 15–25.

Wolfson, J. A., Bleich, S. N., Smith, K. C., & Frattaroli, S. (2016). What does cooking mean to you? Perceptions of cooking and factors related to cooking behaviour. *Appetite, 97*, 146–154.

Wolle, S. (1998). *Die heile Welt der Diktatur: Alltag und Herrschaft in der DDR*. Berlin: Links Verlag.

Index[1]

[1] Note: Page numbers followed by 'n' refer to notes.

P. Jackson et al., *Reframing Convenience Food*,
https://doi.org/10.1007/978-3-319-78151-8

The manufacturer's authorised representative in the EU is Springer Nature Customer Service Centre GmbH, Europaplatz 3, 69115 Heidelberg, Germany. If you have any concerns regarding our products, please contact ProductSafety@springernature.com

Printed and bound by CPI Group (UK) Ltd, Croydon, CR0 4YY
05/10/2026
02228138-0002